Optimizations and Programming

Digital Sciences Set

coordinated by
Abdelkhalak El Hami

Volume 1

Optimizations and Programming

Linear, Nonlinear, Dynamic, Stochastic and Applications with Matlab

Abdelkhalak El Hami
Bouchaib Radi

ISTE

WILEY

First published 2021 in Great Britain and the United States by ISTE Ltd and John Wiley & Sons, Inc.

ISTE Ltd
27-37 St George's Road
London SW19 4EU
UK

www.iste.co.uk

John Wiley & Sons, Inc.
111 River Street
Hoboken, NJ 07030
USA

www.wiley.com

Library of Congress Control Number: 2020948478

British Library Cataloguing-in-Publication Data
A CIP record for this book is available from the British Library
ISBN 978-1-84821-953-3

MIX
Paper from responsible sources
FSC www.fsc.org
FSC® C013604

Contents

Preface

Optimization is the domain of mathematics that studies how to minimize (or maximize) a certain objective, for example, an economic parameter or some type of energy. This is both an ancient topic, with the earliest optimization problems going back to Euclid, and a relatively new topic, with the recent development of numerical methods such as linear programming, which only truly began to flourish in the second half of the 20th century. As an illustration, imagine the problem of searching for a path that joins two points on a map in the shortest time. Some parameters of the problem might be uncertain; for example, we might encounter a traffic jam on the road. Optimization seeks to find the best compromise between the various possible risks.

In the presence of uncertainty, the optimizer (or decider) will often take advantage of information about the system that becomes available dynamically, i.e. gradually over time. For example, we will gradually discover which roads are most susceptible to traffic jams over time, as we try them out. The difficulty of an optimization problem is closely linked to the amount of information needed to make an optimal decision. A problem is said to be large if this quantity of information is too big for classical solving techniques to be applied by brute force.

This book is divided into two parts: programming (Part 1) and optimization (Part 2). In the programming part, we present a collection of tools for operations research, including linear programming, integer programming, binary programming, dynamic programming and stochastic programming. Operations research first arose in the United Kingdom during the Second World War, when scientific methods were used to study various aspects of military operations. Since then, it has become a key element of decision-making processes in various commercial, industrial and governmental contexts, offering a systematic way to apprehend the ever increasing complexity of the management problems faced by both the private and public sectors.

Following its success in military matters during the Second World War, operations research was applied to operational problems in the industry and private sector for many years. In the last decade or so, its applications have been extended to other domains, such as economics, finance, marketing and corporate planning. More recently, operations research has been used to manage healthcare and education systems, solve environmental problems and in other areas of public interest. Its principal users are manufacturing, distribution and retail companies in the mining, energy, transportation and construction sectors, as well as service companies such as banks, and various government bodies. Noteworthy examples of recent applications of operations research include logistical studies, railway safety, packaging design, strategic workforce planning, aerial transportation, forestry operations, nuclear fuel optimization, production planning, and so on.

The optimization part of this book is divided into three chapters: combinatorial optimization, nonlinear optimization without constraints and nonlinear optimization with constraints. It is entirely dedicated to numerical optimization algorithms, their theoretical foundations and convergence properties, their implementation and application, and other practical aspects. The objective is to familiarize readers with these numerical algorithms in order to understand how they behave in practice, how to properly take advantage of *Matlab* as a tool, how to design and adequately implement such algorithms and how to correctly diagnose any difficulties that might arise.

Each chapter starts with a few reminders of key results, but readers should not hesitate to consult the references listed at the end of the book. This book is organized according to a strictly linear approach. As a general rule, the concepts are illustrated with examples. Each chapter ends with an example in *Matlab*.

Acknowledgments

We would like to thank everyone who contributed, directly or indirectly, to the writing of this book and especially, the engineering and PhD students of INSA Rouen whom we have had the pleasure of supervising over the past few years.

Bouchaib RADI
Abdelkhalak EL HAMI
November 2020

PART 1

Programmation

1

Linear Programming

1.1. Introduction

Linear programming can be defined as a mathematical technique for solving management problems. For example, suppose that a manager faced with various options wishes to determine the optimal way to use the resources of a company to achieve a specific objective, such as maximizing the utility or minimizing the cost. The company's problems can usually be modeled as a linear program (LP) consisting of a certain number of resources [HIL 90]. For instance, the labor, raw materials and capital are all resources available in limited quantities that must be distributed optimally over various manufacturing processes. The approach to solving this type of problem is divided into two key steps:

– model the problem with linear equations or inequalities that allow us to properly identify and structure the constraints satisfied by the variables of the model. The contribution of each variable to the company's overarching objective must also be defined as a function that will be optimized;

– find the mathematical optimum using specific linear programming techniques. We will study three methods for solving the various types of linear programming problems. The first is based on graphical solving and is therefore limited to two or three variables. The second method is more algebraic; it motivates the third method presented in this chapter, which is known as the simplex method (or algorithm).

1.2. Definitions

DEFINITION 1.1.– *An LP is in canonical form if it is expressed as follows:*

$$\begin{cases} \max\limits_{x \in \mathbb{R}^n} \ < c, x > \\ s.t. \ \ Ax \leq b \\ \qquad x \geq 0 \end{cases} \qquad [1.1]$$

$A \in \mathcal{M}_{p,n}(\mathbb{R})$, $b \in \mathbb{R}^p$, and $x \in \mathbb{R}^n$. We write $x = (x_1, \ldots, x_n)^T \geq 0$ if $x_1, \ldots, x_n \geq 0$.

An LP is in standard form if it is expressed as follows:

$$\begin{cases} \max\ c^T x \\ \text{s.t.}\ \ Ax = b \\ \qquad\ x \geq 0 \end{cases}$$

Every linear problem can be expressed in canonical form: $\min c^T x = -\max c^T x$.

THEOREM 1.1.– Every LP in standard form can be expressed in canonical form and vice versa.

Terminology

– The function $c^T x$ is the *objective function, cost function* or *loss function.*

– The components of the vector x are called *decision variables.*

– If the vector x satisfies all constraints, then x is an *admissible solution.*

– The set of all admissible solutions is the *admissible set* or *admissible region.*

– An admissible solution x^* that maximizes the loss function (i.e. $c^T x^* \geq c^T x$ for every admissible x) is called an *optimal admissible solution* or *optimal solution.*

EXAMPLE 1.1.– A factory manufactures two products P_1 and P_2 while consuming certain resources: equipment, labor and raw materials. The needs are listed in the following table. Each resource is available in limited quantities.

	P_1	P_2	Availability
Equipment	3	9	81
Labor	4	5	55
Raw materials	2	1	20

The two products P_1 and P_2 yield effective profits of 6 dhs and 4 dhs per unit. The goal is to determine which (possibly non-integer) quantities of the products P_1 and P_2 the factory should produce to maximize the total profit from selling these two products, subject to the availability of the resources.

Let x_1 and x_2 be the quantities of the products P_1 and P_2, respectively.

The LP may be expressed as follows: maximize $z(x_1, x_2) = 6x_1 + 4x_2$ subject to:

– availability of resources:

$$3x_1 + 9x_2 \leq 81$$

$$4x_1 + 5x_2 \leq 55$$

$$2x_1 + x_2 \leq 20$$

– positivity of variables: $x_1,\ x_2 \geq 0$

1.3. Geometry of the linear program

1.3.1. *Polyhedra*

DEFINITION 1.2.– *A polyhedron is a set that can be described as*

$$P = \{x \in \mathbb{R}^n \mid Ax \leq b\},$$

where A is an $m \times n$ matrix and b is a vector in $\mathbb{R}^m$. Note that the admissible set of an LP is a polyhedron.

DEFINITION 1.3.– *A polyhedron P is said to be bounded if there exists a constant c such that $\|x\| \leq c$ for every $x \in P$.*

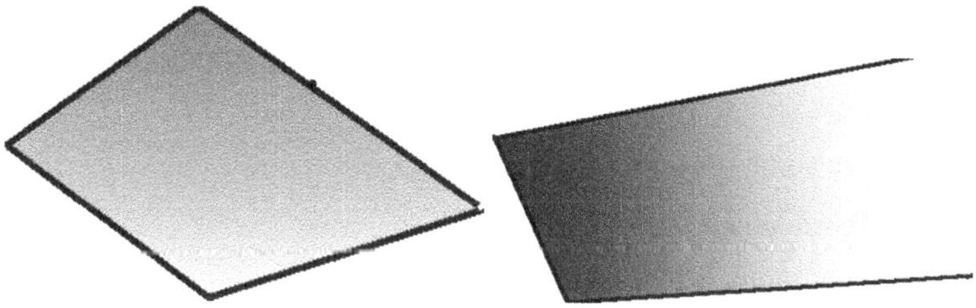

Figure 1.1. *Bounded polyhedron (left) and unbounded polyhedron (right). For a color version of this figure, see www.iste.co.uk/radi/optimizations.zip*

DEFINITION 1.4.– *Let a be a non-zero vector of $\mathbb{R}^n$, and let b be a scalar.*

– The set $\{x \in \mathbb{R}^n | a^T x = b\}$ is said to be a hyperplane.

– The set $\{x \in \mathbb{R}^n | a^T x \geq b\}$ is said to be a half-space.

Note that a hyperplane is the boundary of the corresponding half-space and the vector a is perpendicular to the hyperplane that it defines [TEG 12].

1.3.2. *Extreme points and vertices*

Let P be a polyhedron and $x^* \in P$. The point x^* is an extreme point if and only if x^* is a vertex and x^* is a basic admissible solution.

DEFINITION 1.5.– *Let P be a polyhedron. A vector $x \in P$ is an extreme point of P if it cannot be expressed as a convex combination of two other points of P.*

DEFINITION 1.6.– *Let P be a polyhedron. A vector $x \in P$ is a vertex of P if there exists c such that*

$$c^T x < c^T y$$

for every y in P not equal to x.

THEOREM 1.2.– In linear programming, if the admissible domain is neither empty nor the whole of $\mathbb{R}^n$, it is a convex polytope with finitely many vertices that can either be bounded or unbounded. If an extreme point exists, it is attained at one of the vertices of the polytope. A point in the interior of the domain is never an extreme point if $f \neq 0$. If the polytope is bounded, f attains both a minimum and a maximum on it.

This theorem gives us a graphical solving method.

1.4. Graphical solving of a linear program

The graphical method works by plotting lines and searching for a solution as follows:

– identify the admissible domain;

– identify the contours;

– contours perpendicular to the vector c, and therefore mutually parallel;

– each value of z is associated with a contour;

– the value of z increases in the direction of c.

EXAMPLE 1.2.– Consider, again, the example from earlier. Its mathematical model is defined by the following linear program:

$$\begin{cases} \max\limits_{(x_1,x_2)\in\mathbb{R}^2} & 6x_1 + 4x_2 \\ s.t. & 3x_1 + 9x_2 \leq 81 \\ & 4x_1 + 5x_2 \leq 55 \\ & 2x_1 + x_2 \leq 20 \\ & x_1, x_2 \geq 0 \end{cases}$$

The intersection of the half-planes determined by the lines corresponding to each constraint represents the set of solutions that satisfy the constraints (shaded area). The direction of the cost function is given by

$$z(x_1, x_2) = 6x_1 + 10x_2 = \text{constant}.$$

This gives us the optimal solution $x = \left(\frac{15}{2}, 5\right)$ by sliding the line with this direction upwards, while keeping it parallel, until its intersection with the y-axis is maximal. Graphical solutions are of course only applicable with programs of, at most, three variables (see Figure 1.2).

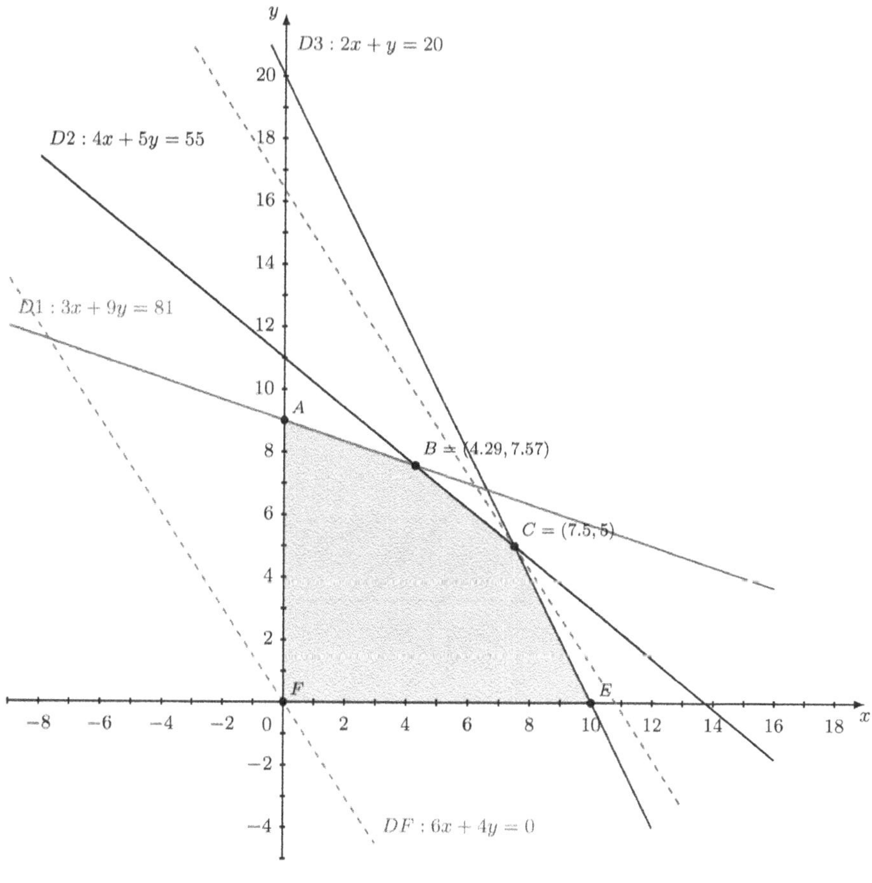

Figure 1.2. *Graphical solution. For a color version of this figure, see www.iste.co.uk/radi/optimizations.zip*

EXAMPLE 1.3.– Consider the linear program:

$$\begin{aligned} \min\ & z = -x_1 - x_2 \\ \text{s.t.}\ & x_1 + 2x_2 \leq 3 \\ & 2x_1 + x_2 \leq 3 \\ & x_1,\ x_2 \geq 0 \end{aligned} \qquad [1.2]$$

The graphical solution is shown in Figure 1.3. We have $z^* = -2$, with $x_1^* = 1$ and $x_2^* = 1$.

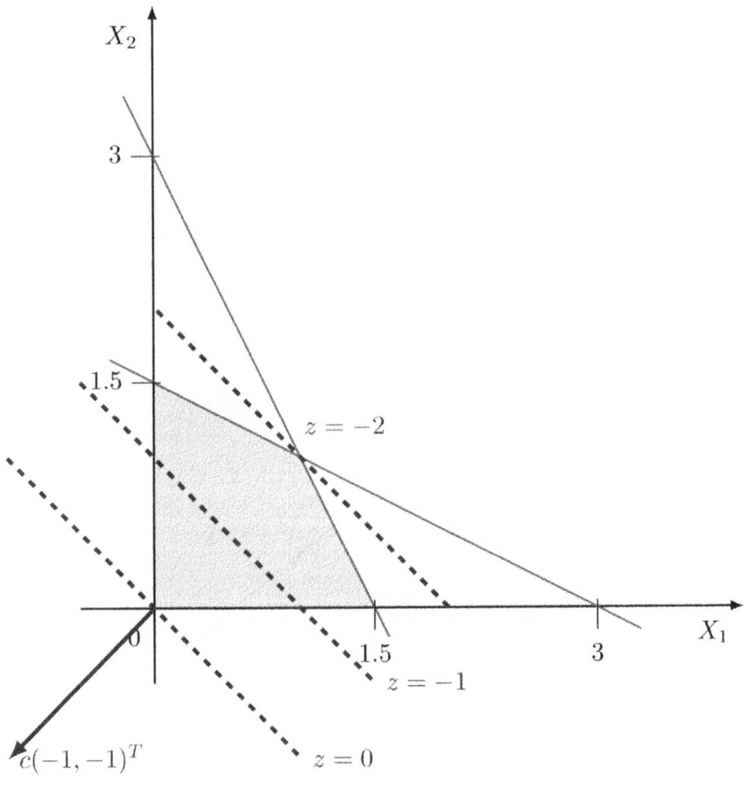

Figure 1.3. *Graphical solution with isoprofit lines. For a color version of this figure, see www.iste.co.uk/radi/optimizations.zip*

1.5. Simplex algorithm

If a linear programming problem in standard form has an optimal solution, then a basic admissible solution that is optimal exists. The simplex algorithm is an algorithm for solving linear optimization problems. Introduced by George Dantzig in 1947, it was probably the earliest algorithm for minimizing functions on sets defined by inequalities. The algorithm moves from one basic admissible solution to the next while reducing the cost.

The idea of the method is as follows:

– Let x_0 be a basic admissible solution.

– For $k = 0, \ldots,$ find an adjacent basic admissible solution x_{k+1} such that $c^T x_{k+1} < c^T x_k$.

– Until no adjacent basic admissible solution improves the objective.

This gives a local minimum, which is in fact a global minimum in linear programming.

1.5.1. *Basic solutions and basic feasible solutions*

Consider a system of m linear equations $Ax = b$ with n variables $(m \leq n)$.

DEFINITION 1.7.– *A basic solution of the system of equations $Ax = b$ is obtained by setting $n - m$ variables to zero and solving the system for the remaining p variables. The solution of the system of p equations in m unknowns is assumed to be unique. The $n - p$ variables set to zero are said to be non-basic variables, and the p remaining variables are said to be basic variables. An LP admits, at most, C_n^m basic solutions. If the basic solution also satisfies the non-negativity constraints, it is said to be a basic feasible solution.*

DEFINITION 1.8.– *A basic solution is said to be degenerate if at least one basic variable is zero. This type of solution is obtained when the number of lines passing through an extreme point is greater than the number of decision variables. A basic feasible solution whose m basic variables are positive is said to be a non-degenerate basic feasible solution.*

REMARK.– Each basic feasible solution corresponds to an extreme point. However, there can be more than one basic feasible solution for the same extreme point. This occurs when the basic feasible solution is degenerate.

REMARK.– The number of basic solutions quickly becomes very large, even in modestly sized models. For example, a model in standard form with 12 constraints and 25 variables can have up to $C_{25}^{12} = 5\,200\,300$ basic solutions.

1.5.2. *Simplex tableau*

Suppose that we wish to solve

$$\begin{cases} \max\ z(x_1,x_2,x_3,x_4) = 4x_1 + 3x_2 \\ \text{s.t.}\ \ 3x_1 + 4x_2 + x_3 = 12 \\ \qquad 7x_1 + 2x_2 + x_4 = 14, \quad x_i \geq 0\ (i = 1, \ldots, 4) \end{cases}$$

Let $c = (4,3,0,0)^T$, $b = (12,14)^T$, $(A, I_2) = \begin{pmatrix} 3 & 4 & 1 & 0 \\ 7 & 2 & 0 & 1 \end{pmatrix}$. It is easy to see that $x = (x_1,x_2,x_3,x_4)^T = (0,0,12,14)^T$ satisfies the constraints. Thus, x is a basic feasible solution with basis $J = \{3,4\}$ (the third and fourth components of x), whose basic variables are $x_3 = 12$ and $x_4 = 14$.

Note that x is not optimal: $z(x) = c^T x = 0$. The optimization procedure constructs a sequence of tables, called simplex tableaus. The first tableau summarizes the data c, b, A' and the basic variables. The last row is equal to $-c^T$.

x_1	x_2	x_3	x_4	
3	4	1	0	$x_1^J = x_3 = 12$
7	2	0	1	$x_2^J = x_4 = 14$
-4	-3	0	0	$z(x) = 0$

Table 1.1. *First simplex tableau*

General case

Consider the problem

$$(P) \begin{cases} \max\ z(x) = c^T x \\ \text{s.t.}\ \ Ax = b, \\ \qquad x \geq 0 \end{cases}$$

where $x \in \mathbb{R}^n_+$, $c \in \mathbb{R}^n$, $b \in \mathbb{R}^m$, and A is an $m \times n$ matrix. We may assume without loss of generality that rank $A = m < n$.

DEFINITION 1.9.– *Let $A = (a^1, a^2, \ldots, a^n)$ (where a^j is the jth column of A), and, for $J \subset \{1,2,\ldots,n\}$, let $A^J = \{a^j,\ j \in J\}$.*

– *$J \subset \{1,2,\ldots,n\}$ is a basis of (P) if $|J| = m$ and if rank A^J = rank $(a^j,\ j \in J) = m$;*

– *let* $x = (x_1, \ldots, x_n)^T \in \mathbb{R}^n_+$. *Then* x_j *is a basic variable (respectively, non-basic variable) if* $j \in J$ *(respectively,* $j \notin J$*). We write* $x^J = (x_j,\ j \in J)$*;*

– *basic feasible solution:* $x = (x_1, \ldots, x_n)^T \in \mathbb{R}^n_+$ *such that* $x_j = 0$ *for* $j \notin J$ *and such that* $Ax = A^J x^J = b$.

REMARK.– The advantage of passing from the canonical form to the standard form is that we immediately obtain a basic feasible solution that can be used as a starting point for the simplex algorithm. The basic variables are the slack variables.

1.5.3. *Change of feasible basis*

Let x be a basic feasible solution of (P). Then

$$z(x) = c^T x = c^J x^J \quad \text{where } c^J = (c_j,\ j \in J).$$

Our goal is to find another feasible basis $\bar{J}$ and a basic feasible solution $\bar{x}$ such that $z(\bar{x}) > z(x)$ (meaning that $\bar{x}$ is better than x). The simplex method proceeds by replacing one of the basic variables x_r with a non-basic variable x_k. We say that

– the variable x_r enters the basis J: $x_r \to \bar{x}_r = 0$;

– the variable x_k leaves the basis $\bar{J}$: $x_k = 0 \to \bar{x}_k > 0$.

Thus, $\bar{J} = (J - \{r\}) \cup \{k\}$. We need rules to choose r and k. These rules are as follows:

– Choose r such that

$$\frac{x_r^J}{a_{rk}} = \min\left\{\frac{x_j^J}{a_{ik}},\ a_{ik} > 0,\ i \in J\right\}. \qquad [1.3]$$

– Choose k such that

$$z_k - c_k = \min_{j=1,\ldots,n}\{z_j - c_j,\ z_j < c_j\ \exists i \in J \mid a_{ij} > 0\}. \qquad [1.4]$$

EXAMPLE 1.4.– Returning to the previous example: $J = \{3, 4\}$, $c_3 = c_4 = 0$, and so $z_j = 0$ for $j = 1, 2, 3, 4$. Thus, $z_j - c_j = -c_j$, $j = 1, 2, 3, 4$. Hence, $k = 1$.

To choose r,

$$\frac{x_r^J}{a_{r1}} = \min\left\{\frac{x_i^J}{a_{i1}}\ ;\ a_{i1} > 0,\ i \in J\right\} = \min\left\{\frac{12}{3}, \frac{14}{7}\right\} = 2.$$

Thus, $r = 2$, x_4 leaves the basis, and x_1 enters the basis. The new basis is $\bar{J} = \{3, 1\}$. We have

$$\bar{x}_1^{\bar{J}} = \bar{x}_3 = x_1^J - \frac{a_{11}}{a_{21}} x_2^J = 12 - \frac{3}{7}14 = 6\bar{x}_2^{\bar{J}} = \bar{x}_1 = \frac{x_2^J}{a_{21}} = 2.$$

Hence, $\bar{x} = (2, 0, 6, 0)^T$ and $z(\bar{x}) = c_1^{\bar{J}}\bar{x}_1^{\bar{J}} + c_2^{\bar{J}}\bar{x}_2^{\bar{J}} = c_3\bar{x}_3 + c_1\bar{x}_1 = 8$. Passing from $J = \{3, 4\}$ to $\bar{J} = \{3, 1\}$ increases the value of z to 8, but this value is not yet maximal.

Calculating the new tableau

We now apply the transformation $x \to \bar{x}$ that increased the value of z. Since the value $z(\bar{x})$ ($> z(x)$) is not necessarily maximal in general, we may need to repeat the steps for choosing r and k several times until we find a basic feasible solution that is also a maximum of z.

To do this, we need a second tableau where x^J is replaced by $\bar{x}^{\bar{J}}$, interpreted in the same way as the first. The same linear transformation that allowed us to pass from x to $\bar{x}$ is applied to the columns of A.

The matrix $A = (a_{ij},\ i = 1, \ldots, m,\ j = 1, \ldots, n)$ is replaced by $\bar{A} = (\bar{a}_{ij},\ i = 1, \ldots, m,\ j = 1, \ldots, n)$ as follows:

$$a^j = \begin{pmatrix} a_{1j} \\ \vdots \\ a_{mj} \end{pmatrix} \to \bar{a}^j = \begin{pmatrix} \bar{a}_{1j} \\ \vdots \\ \bar{a}_{mj} \end{pmatrix}, \text{where } \bar{a}_{ij} = a_{ij} - \frac{a_{ik}a_{rj}}{a_{rk}} (\text{ if } i \neq r) \text{ and } \bar{a}_{rj} = \frac{a_{rj}}{a_{rk}}. \quad [1.5]$$

This gives

$$\sum_{j=1}^{n} a_{ij}x_j = b_i \Leftrightarrow \sum_{j=1}^{n} \bar{a}_{ij}x_j = \bar{b}_i, \quad [1.6]$$

where $\bar{b}_i$, $i = 1, \ldots, m$, are the new basic variables $\bar{x}_i^{\bar{J}}$.

The last row of the new tableau is computed in the same way:

$$\bar{z}_j - c_j = z_j - c_j - \frac{a_{rj}}{a_{rk}}(z_k - c_k). \quad [1.7]$$

EXAMPLE 1.5.– If we apply the above procedure to our example, we obtain Table 1.2.

x_1	x_2	x_3	x_4	
0	22/7	1	-3/7	$x_1^J = x_3 = 6$
1	2/7	0	1/7	$x_2^J = x_1 = 2$
0	-13/7	0	4/7	$z(\bar{x}) = 8$

Table 1.2. *Second simplex tableau ($\bar{J} = \{3, 1\}$)*

As we saw above, the value of z increases from 0 to 8, but there are still negative values in the last row of the tableau, so we need to perform another change of basis by applying the formulas [1.4] and [1.3] after substituting $\bar{J}$. This gives:

$$k = 2,\ \min\left\{\frac{\bar{x}_i^J}{\bar{a}_{i2}}\ ;\ i = 1, 2\right\} = \min\left\{\frac{2}{2/7}, \frac{6}{22/7}\right\} = \min\left\{7, \frac{21}{11}\right\} = \frac{21}{11},$$

$r = 1,\ x_3$ leaves the basis. $\tilde{J} = \{2, 1\}$ is the new basis.

The computation with this new basis is presented in Table 1.3.

x_1	x_2	x_3	x_4	
0	1	7/22	-3/22	$\tilde{x}_1^J = x_2 = 21/11$
1	0	-1/11	2/11	$\tilde{x}_2^J = x_1 = 16/11$
0	0	13/22	7/22	$z(\tilde{x}) = 127/11$

Table 1.3. *Third simplex tableau ($\tilde{J} = \{2, 1\}$)*

The value of z now increases from 8 to $\frac{127}{11}$. This value is maximal because every value in the final row is non-negative. The optimal solution is therefore $x = (x_1, x_2) = (\frac{16}{11}, \frac{21}{11})$. This solution is unique because no further change of basis is possible.

EXAMPLE 1.6.– Consider the linear problem:

$$\begin{aligned}
\min\ & z = -800x_1 - 300x_2 \\
\text{s.t.}\ & 2x_1 + x_2 \leq 400 \\
& x_1 \leq 150 \\
& x_2 \leq 200 \\
& x_1, x_2 \geq 0
\end{aligned} \qquad [1.8]$$

Let us introduce slack variables x_3, x_4 and x_5:

$$\begin{array}{llllllll} \min z = & -800x_1 & - & 300x_2 & & & & \\ \text{s.t.} & 2x_1 & + & x_2 & + x_3 & & & = 400 \\ & x_1 & & & & + x_4 & & = 150 \\ & & & x_2 & & & + x_5 & = 200 \\ & x_1, & & x_2, & x_3, & x_5 & & \geq 0 \end{array} \quad [1.9]$$

This gives the following initial tableau with the basis (x_3, x_4, x_5):

$$T_0 = \begin{array}{|ccccc|c|l} \hline x_1 & x_2 & x_3 & x_4 & x_5 & & \theta \\ \hline 2 & 1 & 1 & 0 & 0 & 400 & 200 \\ 1 & 0 & 0 & 1 & 0 & 150 & 150 \leftarrow \\ 0 & 1 & 0 & 0 & 1 & 200 & \\ \hline -800 & -300 & 0 & 0 & 0 & 0 & \\ \hline \end{array}$$

The new basis is (x_3, x_1, x_5) given as:

$$T_1 = \begin{array}{|ccccc|c|l} \hline x_1 & x_2 & x_3 & x_4 & x_5 & & \theta \\ \hline 0 & 1 & 1 & -2 & 0 & 100 & 100 \leftarrow \\ 1 & 0 & 0 & 1 & 0 & 150 & \\ 0 & 1 & 0 & 0 & 1 & 200 & 200 \\ \hline 0 & -300 & 0 & 800 & 0 & 120000 & \\ \hline \end{array}$$

The next basis is (x_2, x_1, x_5) given as:

$$T_2 = \begin{array}{|ccccc|c|} \hline x_1 & x_2 & x_3 & x_4 & x_5 & \\ \hline 0 & 1 & 1 & -2 & 0 & 100 \\ 1 & 0 & 0 & 1 & 0 & 150 \\ 0 & 0 & -1 & 2 & 1 & 100 \\ \hline 0 & 0 & 300 & 200 & 0 & 150000 \\ \hline \end{array}$$

Since the reduced costs are all positive, this tableau is optimal. The optimal solution is $x_1 = 150$ and $x_2 = 100$, giving an optimal value of $z = -1,500,000$ for the objective function.

1.5.4. *Existence and uniqueness of an optimal solution*

After writing out the first simplex tableau and adding $z_j = -c_j$ to the last row, there are three possible cases:

1) $z_j - c_j \geq 0$ for $j = 1, \ldots, n$. The basic feasible solution x is already optimal. This optimal solution may or may not be unique;

2) there exists $j \in \{1, \ldots, n\}$ such that $z_j - c_j < 0$ and $a_{ij} \leq 0$ for every $i \in J$. In this case, the domain of realizable solutions is unbounded and the problem is ill-posed, since $\max z(x) = +\infty$;

3) the usual case: there exists $j \in \{1, \ldots, n\}$ such that $z_j - c_j < 0$, and there exists $i \in J$ such that $a_{ij} > 0$. The change in basis described earlier is now possible and should be performed, possibly several times, until case (1) is reached.

Could the simplex algorithm ever fail to terminate if case (3) leads to a loop? The answer is yes, and examples have been successfully constructed. However, they are very rare in practice.

Let us therefore state two important theorems about the simplex method.

THEOREM 1.3.– Let $x \in \mathbb{R}^n_+$ be a basic realizable solution of (P) with respect to a basis J ($|J| = m = rank(A)$). Let $z_j = \sum_{i \in J} c_i a_{ij}$. If $z_j - c_j \geq 0$ for every $j = 1, \ldots, n$, then x is an optimal basic realizable solution.

THEOREM 1.4.– Let $x \in \mathbb{R}^n_+$ be a basic realizable solution of (P) with respect to a basis J ($|J| = m = rank(A)$). Let $z_j = \sum_{i \in J} c_i a_{ij}$. Suppose that $a_{ij} \leq 0$ for every $i \in J$ and for every $j \in \{1, \ldots, n\}$ such that $z_j - c_j < 0$. Then the set $\{z(x),\ x \text{ is a realizable solution}\}$ is unbounded.

REMARK.– For a comprehensive discussion of the efficiency of the simplex method, refer to [CHV 83].

1.6. Initialization of the simplex algorithm

This section presents two techniques for finding a basic realizable solution to initialize the simplex algorithm: the first is the Big M method, and the second is the Phase I method.

1.6.1. *Big M method*

Suppose that we wish to solve the LP

$$(P) \quad \begin{aligned} &\max\ z(x) = c^T x \\ &\text{s.t.}\ \ Ax = b, \\ &\qquad x \geq 0 \end{aligned}$$

where A is an $m \times n$ matrix, rank $A = m < n$.

By adding slack variables (with positive or negative signs), we can always reduce the LP to the form (P) described above, with $b \in \mathbb{R}^m_+$. If there is no obvious basic realizable solution to initialize the simplex algorithm, we proceed by adding artificial variables $y_i \geq 0$ to the constraints:

$$Ax = b \text{ is replaced by } Ax + y = b, \ y = (y_1, y_2, \ldots, y_m)^T \in \mathbb{R}^m_+.$$

The new constraints are not equivalent to the initial constraints. The $y_i > 0$ are penalized by replacing the objective function $z(x)$ with

$$z' = c^T x - M \sum_{i=1}^{m} y_i,$$

where M is some very large positive value. We will choose $y_i = b_i, i = 1, \ldots, m$, as a basic feasible solution to serve as the initial solution.

Since the y_i significantly reduce the value of z', they will disappear from the basis over the course of the simplex algorithm. As they are non-basic variables, their value will be equal to zero, and the LP that is ultimately solved is the same as the program (P).

EXAMPLE 1.7.– Consider the LP

$$\begin{aligned} \max \ & z(x) = -x_1 + 2x_2 - 2x_3 \\ \text{s.t.} \ & x_1 + x_2 - x_3 = 3 \\ & -x_1 + 3x_2 = -4 \\ & x_1, x_2, x_3 \geq 0. \end{aligned}$$

The new program is given as

$$\begin{aligned} \max \ & z'(x, y) = (-1 + 2M)x_1 + (2 - 2M)x_2 + (-2 - M)x_3 \\ \text{s.t.} \ & x_1 + x_2 - x_3 + y_1 = 3 \\ & -x_1 + 3x_2 + y_2 = -4 \\ & x_1, x_2, x_3, y_1, y_2 \geq 0. \end{aligned}$$

The first tableau is as follows:

c	-1+2M	2-2M	-2-M	0	0
Basic variables	x_1	x_2	x_3	y_1	y_2
$y_1 = 3$	1	1	-1	1	0
$y_2 = 4$	1	-3	0	0	1
$z'(x) = -7M$	1-2M	-2+2M	2+M	0	0

The second tableau is as follows:

c	-1+2M	2-2M	-2-M	0	0
Basic variables	x_1	x_2	x_3	y_1	y_2
$x_1 = 3$	1	1	-1	1	0
$y_2 = 1$	0	-4	1	-1	1
$z'(x) = -3 - M$	0	-3+4M	3-M	-1+2M	0

The third tableau is as follows:

c	-1+2M	2-2M	-2-M	0	0
Basic variables	x_1	x_2	x_3	y_1	y_2
$x_1 = 4$	1	-3	0	0	1
$x_3 = 1$	0	-4	1	-1	1
$z'(x) = -6$	0	9	0	2+M	-3+M

The unique optimal solution is given by $x_1 = 4$, $x_2 = 0$, $x_3 = 1$, and $z(x_1, x_2, x_3) = -6$

1.6.2. *Auxiliary program or Phase I*

Suppose that the LP that we wish to solve is in the standard form:

$$\begin{cases} \max\ z(x) = c^T x \\ \text{s.t.}\quad Ax = b,\ x \geq 0. \end{cases}$$

The auxiliary program method proceeds by letting $M \to +\infty$ in (P_M) and solving the program obtained in the limit. Let

$$z''(x, y) = \frac{1}{M} z'(x, y) = \frac{c}{M} x - \sum_{i=1}^{m} y_i.$$

Maximizing $z'(x,y)$ is equivalent to maximizing $z''(x,y)$, or maximizing

$$\lim_{M\to+\infty} z''(x,y) = -\sum_{i=1}^{m} y_i.$$

We therefore solve the auxiliary program (P_A) given by

$$(P_A)\begin{cases} \max\ z''(x,y) = -\sum_{i=1}^{m} y_i \\ \text{s.t.}\quad Ax + y = b, \\ \qquad x \in \mathbb{R}_+^n,\ y \in \mathbb{R}_+^m,\ b \in \mathbb{R}_+^m. \end{cases}$$

The simplex algorithm can now be applied to (P_A) starting from the basic realizable solution $y = b$.

– *First case*: $\max z''(y) < 0$. There exists $y_i > 0$, $i \in \{1, 2, \ldots, m\}$. In this case, (P) does not have any realizable solutions.

– *Second case*: $\max z''(y) = 0$. The optimal solution of (P_A) can be used as an initial basic realizable solution for applying the simplex algorithm to (P). The objective function of (P) is expressed in terms of non-basic variables, then the simplex algorithm is applied to (P).

REMARK.– For a minimization problem, we add $\sum y_i$ instead.

Let us go into more detail for a minimization problem. The initial tableau of the auxiliary problem is as follows:

A	b
$-c_B^T A \mid O$	$-c_B^T b$

where $B = I$, $c_B = (1, 1, 1, \ldots)^T$, and the matrices A and b relate to the auxiliary program. For the original variables i, we have $c_i = 0$.

If (x^*, y^*) is a zero-cost optimal solution of the auxiliary problem, then the artificial variable in the basis is necessarily zero, so the solution is degenerate.

If we assume that the kth basic variable is artificial, consider the kth row of the tableau and select the element of this row in column j, where j is the index of a variable in the original problem. If the element is non-zero: k leaves the basis and j enters the basis, and we pivot the tableau. But what if no such element exists? This would mean that the matrix A does not have full rank, and the row in question corresponds to a redundant constraint.

Once the optimal tableau of the auxiliary problem has been obtained and all artificial variables have left the basis, we obtain the initial tableau of the original problem P1 by deleting any columns that relate to the artificial variables and calculating the reduced initial costs: $c^T - c_B^T B^{-1} A$ and $-c_B^T B^{-1} b$ in the box reserved for the cost function. The following example shows how this method works.

EXAMPLE 1.8.– Consider the following linear programming problem:

$$\begin{array}{llcl} \min z = & x_1 + x_2 + x_3 - 2x_4 & & \\ \text{s.c.} & x_1 + x_2 + x_3 & = & 1 \\ & x_2 + x_4 & = & 1 \\ & x_1,\ x_2,\ x_3,\ x_4 & \geq & 0 \end{array} \qquad [1.10]$$

We will first perform Phase I of the simplex algorithm. After adding artificial variables, we obtain the tableau as follows:

$$T_0 = \begin{array}{|cccccc|c|l} \multicolumn{6}{c}{x_1\ x_2\ x_3\ x_4\ x_5\ x_6} & \multicolumn{1}{c}{} & \theta \\ \hline 1 & 1 & 1 & 0 & 1 & 0 & 1 & 1 \leftarrow \\ 0 & 1 & 0 & 1 & 0 & 1 & 1 & \\ \hline -1 & -2 & -1 & -1 & 0 & 0 & -2 & \\ \hline \multicolumn{1}{c}{\uparrow} & & & & & & \multicolumn{1}{c}{} & \end{array}$$

$$T_1 = \begin{array}{|cccccc|c|l} \multicolumn{6}{c}{x_1\ x_2\ x_3\ x_4\ x_5\ x_6} & \multicolumn{1}{c}{} & \theta \\ \hline 1 & 1 & 1 & 0 & 1 & 0 & 1 & 1 \leftarrow \\ 0 & 1 & 0 & 1 & 0 & 1 & 1 & 1 \\ \hline 0 & -1 & 0 & -1 & 1 & 0 & -1 & \\ \hline \multicolumn{1}{c}{} & \uparrow & & & & & \multicolumn{1}{c}{} & \end{array}$$

$$T_2 = \begin{array}{|cccccc|c|l} \multicolumn{6}{c}{x_1\ x_2\ x_3\ x_4\ x_5\ x_6} & \multicolumn{1}{c}{} & \theta \\ \hline 1 & 1 & 1 & 0 & 1 & 0 & 1 & \\ -1 & 0 & -1 & 1 & -1 & 1 & 0 & 0 \leftarrow \\ \hline 1 & 0 & 1 & -1 & 2 & 0 & 0 & \\ \hline \multicolumn{1}{c}{} & & & \uparrow & & & \multicolumn{1}{c}{} & \end{array}$$

$$T_3 = \begin{array}{|cccccc|c|} \multicolumn{6}{c}{x_1\ x_2\ x_3\ x_4\ x_5\ x_6} & \multicolumn{1}{c}{} \\ \hline 1 & 1 & 1 & 0 & 1 & 0 & 1 \\ -1 & 0 & -1 & 1 & -1 & 1 & 0 \\ \hline 0 & 0 & 0 & 0 & 1 & 1 & 0 \\ \hline \end{array}$$

There are no longer any artificial variables in the basis. We have found the optimal tableau of the auxiliary problem, and therefore an admissible basic solution for the initial problem. Since Phase I is now complete, delete the columns corresponding to the artificial variables and compute the reduced costs to obtain the following tableau:

$$T_0 = \begin{array}{|cccc|c|l} \multicolumn{4}{c}{x_1 \; x_2 \; x_3 \; x_4} & & \theta \\ \hline 1 & 1 & 1 & 0 & 1 & 1 \leftarrow \\ -1 & 0 & -1 & 1 & 0 & \\ \hline -2 & 0 & -2 & 0 & -1 & \\ \hline \multicolumn{4}{l}{\uparrow} & & \end{array}$$

$$T_1 = \begin{array}{|cccc|c|} \multicolumn{4}{c}{x_1 \; x_2 \; x_3 \; x_4} & \\ \hline 1 & 1 & 1 & 0 & 1 \\ 0 & 1 & 0 & 1 & 1 \\ \hline 0 & 2 & 0 & 0 & 1 \\ \hline \end{array}$$

This tableau is optimal. The optimal solution is $x_1 = 1$, $x_2 = 0$, $x_3 = 0$ and $x_4 = 1$, giving an optimal value of $z = -1$.

1.6.3. *Degeneracy and cycling*

Suppose that the current basic admissible solution is degenerate (i.e. at least one basic variable is zero), and let us recall a possible unfavorable scenario that can occur when the simplex algorithm is iterated. Performing a change of basis yields a value for the cost function that is greater than or equal to the previous value. The method is only guaranteed to converge if $z^* < z_0$; in fact, if the previous basic solution is degenerate, it might be the case that $z^* = z_0$.

If this phenomenon occurs multiple times over consecutive iterations, the algorithm can return to a basis that was already considered earlier. In other words, it returns to the same vertex twice. This is called cycling.

Although degeneracy is a frequently encountered phenomenon in linear programming, cycling is not. Some examples have been given in the literature [GAS 75]. Despite the rarity of cycling, it would be desirable to specify a method that avoids it and therefore guarantees that the simplex method will converge. There are several possible choices for a rule that avoids the inconvenience of cycling.

In the past, the most common technique was the so-called method of small perturbations. This method applies an infinitesimal geometric modification to the vector d in order to force it to be expressible as a linear combination of m vectors, with strictly positive coefficients in some basis. Each vertex is then defined as the intersection of n hyperplanes [HIL 90].

More recently, Bland [BLA 77] suggested modifying the rules for choosing the change of basis. Bland's rule proceeds as follows:

– for the variable entering the basis, choose the smallest index j such that the reduced cost is negative;

– for the variable leaving the basis, if there is equality in θ^*, choose the variable with the lowest index.

1.6.4. *Geometric structure of realizable solutions*

Earlier, when we solved linear programs graphically, the optimal solutions were on the boundary of the convex set of realizable solutions. If there was a unique optimal solution, it was an extreme point.

Recall the canonical form of a linear program:

$$(P)\begin{cases} \max\ z(x) = c^T x \\ \text{s.t.} \quad Ax \leq b, \\ \qquad\quad x \geq 0, \end{cases}$$

where the matrix A is assumed to be an $m \times n$ matrix of rank m $(< n)$. The standard form is then given by

$$(P_s)\begin{cases} \max \qquad z(x) = c^T x \\ s.t.\ \ (A, I_m)\begin{pmatrix} x \\ e \end{pmatrix} = b, \\ \qquad\quad x \in \mathbb{R}^n_+,\ e \in \mathbb{R}^m_+ \end{cases}$$

It is possible to show the following result.

Let Γ (respectively, $\sum$) be the set of feasible solutions of (P) (respectively of (P_s)):

$$\Gamma = \{x \in \mathbb{R}^n\ ;\ Ax \leq b,\ x \geq 0\},$$

$$\sum = \left\{\begin{pmatrix} x \\ e \end{pmatrix},\ (A, I_m)\begin{pmatrix} x \\ e \end{pmatrix} = 0,\ x \in \mathbb{R}^n_+,\ e \in \mathbb{R}^m_+\right\}.$$

Then Γ and $\sum$ are closed and convex sets.

Recall that the simplex method only uses the basic solutions of (P_s). The next theorem tells us that these are in fact the extreme points of $\sum$.

THEOREM 1.5.– $\begin{pmatrix} x^0 \\ e^0 \end{pmatrix}$ is a basic feasible solution of (P_s) if and only if $\begin{pmatrix} x^0 \\ e^0 \end{pmatrix}$ is an extreme point of $\sum$.

Examples where the solution is not unique show that the optimal solutions lie on the boundary:

THEOREM 1.6.– Every optimal solution of (P) (respectively of (P_s)) belongs to the boundary of Γ (respectively of $\sum$).

A question presents itself: can there be optimal solutions but no optimal basic feasible solutions? If so, the simplex method, which only considers basic solutions, would not be able to find an optimal solution. The following theorem tells us that this does not happen.

THEOREM 1.7.– If (P_s) has an optimal solution, then (P_s) has an optimal basic feasible solution.

1.7. Interior-point algorithm

The simplex algorithm moves along a sequence of vertices of the polyhedron and converges rapidly on average. However, Klee and Minty [KLE 72] found pathological examples where the simplex algorithm visits almost every vertex, which can potentially be an enormous number. The existence of theoretical linear programs that derail the simplex algorithm motivated research to find an algorithm that has polynomial complexity in m and n, even in the worst-case scenarios. Khachian [KHA 79] found the first such algorithm, based on the ellipsoid method from nonlinear optimization. Although it is faster than the simplex method in Minty's problems, it never succeeded in establishing itself because it tended to be much slower than the simplex algorithm on real problems in practice. Nevertheless, as a mathematical result, it inspired research into interior-point algorithms.

In 1984, Karmarkar [KAR 84] found a new algorithm. The computation time in the worst-case scenario is proportional to $n^{3.4}L$, where L denotes the number of bits required to encode the tableaus A, b and c that define the linear program. The algorithm is too complicated for computations by hand; known as the interior-point algorithm, it enters into the polyhedron itself to converge on the optimal vertex more quickly.

Today, some variants of this algorithm are able to compete with the simplex algorithm. Some software programs already include an interior-point algorithm in addition to the simplex algorithm. On some types of linear programs, interior-point algorithms can be 10 times faster than the simplex algorithm, but nevertheless, there are still many cases where the simplex algorithm is quicker [PRI 11].

EXAMPLE 1.9.– Consider the LP

$$\begin{aligned} \max\ & z = 2x_1 + x_2 \\ \text{s.t.}\ & x_1 + x_2 \leq 4 \\ & -x_1 + x_2 \leq 2 \\ & x_1 \leq 2 \\ & 0.5x_1 - x_2 \leq 0 \\ & x_1, x_2 \geq 0 \end{aligned}$$

The optimum is $x^* = (2,2)^T$ with cost $z^* = 6$. Starting from the interior point $x_0 = (1,1)^T$ and without a safety factor, the algorithm performs two iterations: it considers the point $x_1 = (2, 1.591)^T$, followed by the point x^*.

1.8. Duality

Duality is an essential notion in linear programming. The duality operation associates any so-called primal linear programming problem with another such problem, known as the dual, that is defined solely in terms of the data of the primal problem.

Duality is interesting for two reasons:

– the dual problem has an important economic interpretation that offers another perspective of the primal problem;

– there are straightforward and powerful mathematical relations between the primal and dual problems; this will allow us to develop new algorithms that will be more efficient in many situations.

Suppose that we have the following primal program:

$$(P) \begin{cases} \max\ z(x) = cx \\ \text{s.t.} \quad Ax \leq b, \\ \qquad\quad x \in \mathbb{R}^n_+ \end{cases}$$

Its dual program is as follows:

$$(D)\begin{cases}\min\ w(y) = yb\\ \text{s.t.}\quad yA \geq c,\\ \qquad\quad y \in \mathbb{R}^m_+\end{cases}$$

REMARK.– The primal and dual programs satisfy the following relations:

– m = number of constraints of (P) = number of variables of (D),

– n = number of variables of (P) = number of constraints of (D).

If (P) has two constraints, (D) has two variables and can be solved graphically, regardless of the number of variables of (P).

So, what is the form of (D) when (P) is not in canonical form?

In such a case, we can determine (D) by reducing (P) to canonical form, as shown in the following example.

EXAMPLE 1.10.– Consider the LP:

$$(P)\begin{cases}\max\ z(x) = c^T x\\ \text{s.t.}\quad Ax = b,\ x \geq 0 \Leftrightarrow \begin{pmatrix} A\\ -A\end{pmatrix} x \leq \begin{pmatrix} b\\ -b\end{pmatrix},\ x \geq 0\end{cases}$$

By the above definition, the dual of (P) is given by

$$(D)\begin{cases}\min\ w(u,v) = (u,v)\begin{pmatrix} b\\ -b\end{pmatrix} = (u-v)b\\ \text{s.t.}\quad (u,v)\begin{pmatrix} A\\ -A\end{pmatrix} = (u-v)A \geq c,\ u,v \in \mathbb{R}^m_+.\end{cases}$$

Setting $y = u - v$ gives

(D) $\min\ w(y) = yb$ s.t. $yA \geq c$, y with no sign restriction (denoted (y)).

In general, we have:

Primal	Dual
Maximization	Minimization
Coefficient of z	Right-hand side of the constraints
Right-hand side of the constraints	Coefficient of w

$$\text{Constraint}\begin{cases} = \\ \leq \\ \geq \end{cases} \qquad \text{Variable}\begin{cases} \text{no sign constraint} \\ \leq \\ \geq \end{cases}$$

$$\text{Variable}\begin{cases} \text{no sign constraint} \\ \geq \\ \leq \end{cases} \qquad \text{Constraint}\begin{cases} = \\ \leq \\ \geq \end{cases}$$

THEOREM 1.8.– The dual of the dual is the primal.

1.8.1. *Duality theorem*

Now that we have defined the dual of a linear program, let us study the links between the solutions of programs that are dual to one another.

THEOREM 1.9.– Let (x, y) be a feasible solution of $(\bar{P})$ and (u, v) a feasible solution of $(\bar{D})$. Then:

1) $z(x, y) \leq w(u, v)$,

2) $z(x, y) = w(u, v) \Rightarrow (x, y)$ and (u, v) are optimal solutions of $(\bar{P})$ and $(\bar{D})$.

THEOREM 1.10.– One of the following three cases is satisfied:

1) $(\bar{P})$ and $(\bar{D})$ have optimal solutions and $\max z(x, y) = \min w(u, v)$;

2) one of $(\bar{P})$ and $(\bar{D})$ has a feasible solution, but not the other;

3) neither $(\bar{P})$ nor $(\bar{D})$ have feasible solutions.

THEOREM 1.11.– Let (x, y) (respectively, (u, v)) be feasible solutions of $(\bar{P})$ (respectively, $(\bar{D})$). Then (x, y) and (u, v) are optimal solutions of $(\bar{P})$ and $(\bar{D})$ if and only if the following statements hold:

– if one constraint is strictly an inequality in $(\bar{P})$ (respectively, $(\bar{D})$), then the corresponding variable of $(\bar{D})$ (respectively, of $(\bar{P})$) is zero;

– if the value of a restricted variable (i.e. a positive variable or a negative variable) in one of the programs $(\bar{P})$ or $(\bar{D})$ is not zero, then the corresponding constraint in the other program is an equality.

This theorem can alternatively be stated as follows:

THEOREM 1.12.– Let x and p be admissible solutions of the primal and dual programs, respectively. Then the vectors x and p are, respectively, optimal solutions of the two problems if and only if:

$$\begin{aligned} p_i(a_i^T x - b_i) &= 0 \ \forall i \\ (c_j - p^T A_j)x_j &= 0 \ \forall j \end{aligned}$$

Application

Consider the LP

$$\begin{aligned} \min\ & 13x_1 + 10x_2 + 6x_3 \\ \text{s.t.}\ & 5x_1 + x_2 + 3x_3 = 8 \\ & 3x_1 + x_2 = 3 \\ & x_i \geq 0 \end{aligned}$$

The dual of this program is:

$$\begin{aligned} \max\ & 8p_1 + 3p_2 \\ \text{s.t.}\ & 5p_1 + 3p_2 \leq 13 \\ & p_1 + p_2 \leq 10 \\ & 3p_1 \leq 6 \end{aligned}$$

The solution of the primal is $x^* = (1, 0, 1)^T$.

The optimal dual solution can be constructed from additional slack conditions:

– the condition $p_i(a_i^T x^* - b_i) = 0$ is satisfied because x^* is admissible;

– the condition $(c_j - p^T A_j)x_j = 0$ is satisfied for $j = 2$:

- for $j = 1$, this condition becomes:

$$5p_1 + 3p_2 = 13.$$

- for $j = 3$, it becomes:

$3p_1 = 6.$

These two conditions give $p_1 = 2$ and $p_2 = 1$. This solution is dual admissible. The total dual cost is 19, and so is the primal cost.

1.9. Relaxation

In mathematical programming, good performance essentially depends on the program's ability to construct useful bounds (lower bounds for maximization and upper bounds for minimization). These bounds can be obtained by relaxing the linear program, i.e.

– either increasing the set of solutions to optimize over a larger space (that is easier to optimize); this includes constraint relaxation (and in particular continuous relaxation);

– or replacing the objective function with an upper bound (in the case of maximization), for example, Lagrangian relaxation.

For a mathematical program, the following continuous relaxations are possible:

– (simple) continuous maximization, which is very effective for LPs (with the simplex algorithm, the interior-point method, or the volume algorithm); however, we will see that continuous relaxation is rarely the best option;

– in the specific context of LPs, we can improve the value of linear relaxation by adding constraints – reinforcement;

– we can also add variables – reformulation;

– Lagrangian relaxation is another approach.

1.9.1. *Lagrangian relaxation*

Lagrangian duality provides a particularly fruitful framework for relaxation. It works as follows:

– choose an inequality in the LP and remove it from the LP;

– add this inequality to the objective function with a multiplier that penalizes the objective function if the solution found does not satisfy the inequality.

This relaxation often produces a very good relaxation value, far better than continuous relaxation.

Consider the following LP:

$$\begin{aligned} &\min\ f(x) \\ &\text{s.t.}\ \ g_i(x) \leq 0 \quad i \in I \\ &\qquad x \in K \subset \mathbb{R}^n \end{aligned} \qquad [1.11]$$

To each constraint $i \in I$, we assign a real number $\lambda_i \geq 0$ called the Lagrange multiplier. We also define the vector $\lambda = (\lambda_i, 0\ i \in I)$. The Lagrangian function is then defined as

$$\mathcal{L}(x, \lambda) = f(x) + \sum_{i \in I} \lambda_i gi(x).$$

This function has the two vector arguments x and λ and takes values in $\mathbb{R}$. Suppose that the PL is such that the problem $\min_{x \in S} \mathcal{L}(x, \lambda)$ has an optimal solution for every $\lambda \geq 0$.

More details are given in Chapter 7. Now consider the following LP [FLE 10]:

$$\begin{aligned} &\min\ z = 4x_1 + 2x_2 + x_3 \\ &\text{s.t.}\ \ x_1 + x_2 \geq 1.1 \\ &\qquad x_2 + x_3 \geq 3.2 \\ &\qquad x_1 + x_3 \geq 4.4 \\ &\qquad x_i \geq 0 \end{aligned} \qquad [1.12]$$

We obtain the optimal solution $(0, 1.1, 4.4)^T$, which gives the optimal cost as $z^* = 6.6$.

By relaxing the third constraint ($x_1 + x_2 \geq 4.4$) and keeping the integrality constraints on the variables x_i, we obtain the following LP by adding the expression $\lambda(4.4 - x_1 - x_3)$ to the objective function, where λ is the Lagrangian relaxation coefficient, which is updated at each iteration of the method:

$$\begin{aligned} &\min\ f_\lambda(x_i) = 4x_1 + 2x_2 + x_3 + \lambda(4.4 - x_1 - x_3) \\ &\text{s.t.}\ \ x_1 + x_2 \geq 1.1 \\ &\qquad x_2 + x_3 \geq 3.2 \\ &\qquad x_i \geq 0 \\ &\qquad x_i \leq 10 \end{aligned} \qquad [1.13]$$

The last constraint is added to bound the variables x_i. This guarantees that the problem is bounded for every $\lambda > 0$. The coefficient λ is updated as follows: $\lambda^{k+1} = \max\left(0, \lambda^k + \frac{1}{k} \times (4.4 - x_1 - x_3)\right)$. It can be shown that λ tends to 1 and z tends to 8.4.

1.10. Postoptimal analysis

Consider the LP:

$$\begin{aligned} \min\ & z = c^T x \\ \text{s.t.}\ & Ax \leq b \\ & x \geq 0 \end{aligned} \qquad [1.14]$$

We will now perform a postoptimal analysis of the optimal solution in terms of the problem data, also known as a sensitivity analysis. In other words, how does the solution or the objective function vary if we modify one of the entries of the vectors c or b, or the matrix A?

To conduct a postoptimal analysis, we need to be able to express the solution of the simplex algorithm in terms of the initial problem data instead of simplex tableaus, which change at each iteration.

As always, we introduce slack variables to the problem [1.14], writing A for the resulting matrix of size $m \times (m+n)$. We also assume that the basic variables B of the optimal solution x_B are known. This information can be read off from the final tableau of the simplex algorithm, which is why this procedure is called postoptimal analysis.

Let $B = \{x_{j_1}, x_{j_2}, \ldots, x_{j_m}\}$ be the list of basic variables, and let N be its complement with respect to the index set $\{1, 2, \ldots, m+n\}$. For example, if $m = n = 3$ and $B = \{x_1, x_5, x_2\}$, then $N = \{x_3, x_4, x_6\}$. Based on the sets B and N, we can extract the following submatrices of A:

– A_B is the submatrix formed by the columns corresponding to the variables of B. We take them in the order specified by B.

– A_N is the submatrix formed by the columns corresponding to the variables of N in the order specified by N.

EXAMPLE 1.11.– Consider the problem

$$\begin{aligned} \min\ & z = -1,100x_1 - 1,400x_2 - 1,500x_3 \\ \text{s.t.}\ & x_1 + x_2 + x_3 \leq 340 \\ & 2x_1 + 3x_2 + x_3 \leq 2,400 \\ & x_1 + 2x_2 + 3x_3 \leq 560 \\ & x_1, x_2, x_3 \geq 0 \end{aligned} \qquad [1.15]$$

The matrix system $Ax = b$ with slack variables is given as

$$\begin{bmatrix} 1 & 1 & 1 & 1 & 0 & 0 \\ 2 & 3 & 1 & 0 & 1 & 0 \\ 1 & 2 & 3 & 0 & 0 & 1 \end{bmatrix} \begin{Bmatrix} x_1 \\ x_2 \\ \vdots \\ x_6 \end{Bmatrix} = \begin{Bmatrix} 340 \\ 2,400 \\ 560 \end{Bmatrix}.$$

The final tableau of the simplex algorithm is written as

x_1	x_2	x_3	x_4	x_5	x_6	
1	0	-1	2	0	-1	120
0	0	-3	-1	1	-1	1,500
0	1	2	-1	0	1	220
0	0	200	800	0	300	440

We therefore indeed have $B = \{x_1, x_5, x_2\}$ and $N = \{x_3, x_4, x_6\}$. The submatrices A_B and A_N can be written as:

$$A_B = \begin{bmatrix} 1 & 0 & 1 \\ 2 & 1 & 3 \\ 1 & 0 & 2 \end{bmatrix} \quad A_N = \begin{bmatrix} 1 & 1 & 0 \\ 1 & 0 & 0 \\ 3 & 0 & 1 \end{bmatrix}.$$

Using a block partition based on the index sets B and N, we can write

$$Ax = b \leftrightarrow [A_B \; A_N] \begin{bmatrix} x_B \\ x_N \end{bmatrix} = b \leftrightarrow A_B x_B + A_N x_N = b.$$

The solution $x_B : x_B = A_B^{-1}(b - A_N x_N)$ can be computed algebraically. The matrix A_B is invertible because B is a basis. The optimal solution is given by the formula $x_B = A_B^{-1} b$, since the non-basic variables must be zero, i.e. $x_N = 0$. In our example, we have

$$x_B = \begin{Bmatrix} x_1 \\ x_5 \\ x_2 \end{Bmatrix} = \begin{bmatrix} 2 & 0 & -1 \\ -1 & 1 & -1 \\ -1 & 0 & 1 \end{bmatrix} \begin{Bmatrix} 340 \\ 2,400 \\ 560 \end{Bmatrix} = \begin{Bmatrix} 120 \\ 1,500 \\ 220 \end{Bmatrix},$$

which is the correct optimal solution according to the final simplex tableau. The other columns of the final tableau correspond to

$$A_B^{-1} A_N = \begin{bmatrix} 2 & 0 & -1 \\ -1 & 1 & -1 \\ -1 & 0 & 1 \end{bmatrix} \begin{bmatrix} 1 & 1 & 0 \\ 1 & 0 & 0 \\ 3 & 0 & 1 \end{bmatrix} = \begin{bmatrix} -1 & 2 & -1 \\ -3 & -1 & -1 \\ 2 & -1 & 1 \end{bmatrix},$$

which are indeed the columns that correspond to $N = \{x_3, x_4, x_6\}$.

1.10.1. *Effect of modifying* b

Let us analyze the effect of modifying the vector b. In other words, we shall study the behavior of the solution of the modified problem when b is replaced by $\tilde{b} = b + \Delta b$.

$$\begin{aligned} \min\ & z = c^T x \\ \text{s.t.}\ & Ax \leq \tilde{b} \\ & x \geq 0 \end{aligned} \qquad [1.16]$$

Let x_B be the basic variables of the solution. Our goal is to determine a condition that guarantees that the basis B will remain optimal. In fact, this is easy. The vector b only appears in the optimality condition [1.16]. Therefore, the basic variables x_B remain optimal for the modified problem if

$$A_B^{-1}\tilde{b} \geq 0 \leftrightarrow A_B^{-1}\Delta b \geq -A_B^{-1}b. \qquad [1.17]$$

EXAMPLE 1.12.– Consider the LP:

$$\min \quad z = x_1 + \tfrac{3}{2}x_2 + 3x_3 \qquad [1.18]$$

$$\text{s.t.} \begin{cases} x_1 + x_2 + 2x_3 \geq 6 \\ x_1 + 2x_2 + x_3 \geq 10 \\ x_1, x_2, x_3 \geq 0 \end{cases} \qquad [1.19]$$

In matrix form with slack variables, this can be written as:

$$A = \begin{bmatrix} -1 & -1 & -2 & 1 & 0 \\ -1 & -2 & -1 & 0 & 1 \end{bmatrix}, \quad b = \begin{Bmatrix} -6 \\ -10 \end{Bmatrix}, \quad c = \begin{Bmatrix} 1 \\ 3/2 \\ 3 \\ 0 \\ 0 \end{Bmatrix}.$$

Suppose that the optimal basis is $B = \{x_1, x_2\}$. Then

$$A_B = \begin{bmatrix} -1 & -1 \\ -1 & -2 \end{bmatrix} \Rightarrow A_B^{-1} = \begin{bmatrix} -2 & 1 \\ 1 & -1 \end{bmatrix}.$$

Compute x_B:

$$x_B = A_B^{-1}b = \begin{bmatrix} -2 & 1 \\ 1 & -1 \end{bmatrix} \begin{Bmatrix} -6 \\ -10 \end{Bmatrix} = \begin{Bmatrix} 2 \\ 4 \end{Bmatrix} \geq 0.$$

Compute the reduced costs $c_N^T - c_B^T A_B^{-1} A_N$:

$$c_N^T - c_B^T A_B^{-1} A_N = [\frac{3}{2}, \frac{1}{2}, \frac{1}{2}] \geq 0.$$

The solution $x_B = (2,4)^T$ is therefore optimal, i.e. $x = (2,4,0,0,0)^T$.

If $-b \mapsto -\tilde{b} = \begin{Bmatrix} -6-a \\ -10 \end{Bmatrix}$, i.e. if we consider a change in b_1 only, then:

$$\tilde{x}_B = A_B^{-1}(-\tilde{b}) = \begin{bmatrix} -2 & 1 \\ 1 & -1 \end{bmatrix} \begin{Bmatrix} -6-a \\ -10 \end{Bmatrix} = \begin{Bmatrix} 2+2a \\ 4-a \end{Bmatrix}.$$

We will have $\tilde{x}_B \geq 0$ if and only if

$$2 + 2a \geq 0 \quad \text{and} \quad 4 - a \geq 0.$$

This gives an interval for the parameter b_1: $-1 \leq a \leq 4$.

What is the interval for b_2?

If $-b \mapsto -\tilde{b} = \begin{Bmatrix} -6 \\ -10-a \end{Bmatrix}$, then:

$$\tilde{x}_B = A_B^{-1}(-\tilde{b}) = \begin{bmatrix} -2 & 1 \\ 1 & -1 \end{bmatrix} \begin{Bmatrix} -6 \\ -10-a \end{Bmatrix} = \begin{Bmatrix} 2-a \\ 4+a \end{Bmatrix}.$$

We will have $\tilde{x}_B \geq 0$ if and only if

$$2 - a \geq 0 \quad \text{and} \quad 4 + a \geq 0.$$

This gives the interval for the parameter b_2: $-4 \leq a \leq 2$.

What is the effect on the minimum value of the objective function?

The effect is:

$$z(\tilde{b}) = c_B^T \tilde{x}_B + c_N^T \tilde{x}_N = c_B^T \tilde{x}_B = [1, \frac{3}{2}] \begin{Bmatrix} 2-a \\ 4+a \end{Bmatrix} = 8 + \frac{a}{2}.$$

1.10.2. *Effect of modifying* c

Next, let us find a condition that guarantees that the basis B will remain optimal under $c \mapsto \tilde{c} = c + \Delta c$. First, observe that the solution $x_B = A_B^{-1} b$ remains unchanged. However, the simplex tableau might no longer be optimal. To keep it optimal, we must impose the stability condition

$$\tilde{c}_N^T - \tilde{c}_B^T A_B^{-1} A_N \geq 0 \leftrightarrow (c + \Delta x)_N^T - (c - \Delta c)_B^T A_B^{-1} A_N \geq 0. \qquad [1.20]$$

In general, let us determine the effect of modifying a single variable x_i:

$$c_i \mapsto \tilde{c}_i = c_i + \Delta c_i$$

There are two cases to analyze depending on whether x_i is a basic or non-basic variable.

Case of a non-basic variable

Write e_i for the canonical basis vector of $\mathbb{R}^n$, i.e. $e_i = (0, 0, \ldots, 1, 0, \ldots, 0)^T$ with 1 at position i. Then:

$$\begin{aligned} \Delta c &= \Delta c_i e_i \\ (\Delta c)_B &= 0 \\ (\Delta c)_N &= \Delta c_i e_i \end{aligned}$$

The ith component of the vector $\tilde{c}_N^T - \tilde{c}_B^T A_B^{-1} A_N$ is

$$\begin{aligned} &c_i + \Delta c_i - (c_B^T A_B^{-1} A_N)_i \geq 0 \\ &\Delta c_i \geq -(c_i - (c_B^T A_B^{-1} A_N)_i) = -r_i \ , \\ &\tilde{c}_i \geq c_i - r_i \end{aligned}$$

where the r_i are the components of the final row of the simplex tableau. In other words, we have all the information that we need to compute the stability intervals of the coefficients c_i.

EXAMPLE 1.13.– Again, consider the previous example. The final simplex tableau is:

x_1	x_2	x_3	x_4	x_5	
1	0	3	-2	1	2
0	1	-1	1	-1	4
0	0	3/2	1/2	1/2	-8

The last row gives us the vector of reduced costs $r = (0, 0, \frac{3}{2}, \frac{1}{2}, \frac{1}{2})^T$. The basic variables are $B = \{x_1, x_2\}$, and the non-basic variables are $N = \{x_3, x_4, x_5\}$. The objective function is of the form $z = c_1x_1 + c_2x_2 + c_3x_3$ with $c_1 = 1$, $c_2 = \frac{3}{2}$, and $c_3 = 3$. It does not make sense to perturb the coefficients c_4 and c_5. Accordingly, we can only perturb x_3. Let us compute the stability interval around c_3 associated with the non-basic variable x_3. The stability condition is

$$\tilde{c}_3 \geq c_3 - r_3 = \frac{3}{2} \leftrightarrow \tilde{c}_3 \geq \frac{3}{2}.$$

The optimal solution $x = (2, 4, 0, 0, 0)^T$ is the same for every value $\tilde{c}_3 \geq \frac{3}{2}$. Furthermore, the value of the objective function ($z = 8$) remains unchanged because the variable is non-basic.

1.11. Application to an inventory problem

Consider the problem of an automobile manufacturer that offers two models for sale, a small automobile and a large automobile. Suppose that there is sufficient demand that the manufacturer is certain to sell whatever is produced at the current sales price of 16,000 euros for large automobiles, and 10,000 euros for small ones. The manufacturer's only problem is the limited supply of two raw materials, rubber and steel. Manufacturing a small automobile requires one unit of rubber and one unit of steel, whereas manufacturing a large automobile requires one unit of rubber and two units of steel. If the manufacturer has 400 units of rubber and 600 units of steel in stock, how many small and large vehicles should be produced from this inventory to maximize the turnover?

Let x be the number of large automobiles manufactured, y the number of small automobiles manufactured and z the resulting turnover. The problem can therefore be expressed in the form:

$$\begin{aligned} \max\ z = &\ 16,000x + 10,000y \\ \text{s.t.}\quad & x + y \leq 400 \\ & 2x + y \leq 600 \\ & x \geq 0,\ y \geq 0 \end{aligned} \qquad [1.21]$$

1.11.1. *Optimal solution*

It is relatively easy to find a graphical solution for systems like this with only two variables. This allows us to find the optimal solution $x = 200$ and $y = 200$, which corresponds to $z = 5,200,000$. In this particular example, the optimal solution is unique, corresponding to a vertex of the region bounded by the constraints.

1.11.2. *Sensitivity to variation in stock*

Let us observe how the solution of the problem evolves when some of the starting data is modified, for example by increasing the stock of rubber or steel. Suppose that 700 units of steel are in stock instead of 600. The problem now becomes:

$$\begin{aligned} \max\ z &= 16,000x + 10,000y \\ \text{s.t.} \quad & x + y \leq 400 \\ & 2x + y \leq 700 \\ & x \geq 0,\ y \geq 0 \end{aligned} \qquad [1.22]$$

By solving graphically again, we find that the optimal solution is now $x = 300$ and $y = 100$, giving $z = 5,800,000$. In other words, increasing the available steel by 100 units increases the turnover by 600,000 euros. We say that the marginal price of a unit of steel is 6,000 euros.

If the amount of steel in stock is increased to 800, the optimal solution becomes $x = 400$ and $y = 0$, and the turnover becomes $z = 6,400,000$. Increasing the steel in stock beyond 800 without also changing the rubber in stock no longer affects the optimal solution, because y is constrained to remain positive.

Suppose now that the steel in stock remains fixed at 600, but the rubber in stock is increased by 400 to 500. The new problem is

$$\begin{aligned} \max\ z &= 16,000x + 10,000y \\ \text{s.t.} \quad & x + y \leq 500 \\ & 2x + y \leq 600 \\ & x \geq 0,\ y \geq 0 \end{aligned} \qquad [1.23]$$

Graphical solving shows that the optimal solution is now given by $x = 100$ and $y = 400$, which corresponds to $z = 5,600,000$. In other words, increasing the rubber by 100 units changes the turnover by 400,000 euros. We say that the marginal price of a unit of rubber is 4,000 euros.

If the rubber in stock is increased to 600, the optimal solution becomes $x = 0$ and $y = 600$, and the turnover becomes $z = 6,000,000$. Increasing the amount of rubber in stock beyond 600, without also increasing the amount of steel in stock, no longer affects the optimal solution because x is constrained to remain positive.

1.11.3. *Dual problem of the competitor*

Suppose now that the automobile manufacturer has a competitor who offers to repurchase all raw materials in stock to fulfill a large number of orders. The competitor must offer a certain price (assumed to be constant and denoted u) per unit of rubber and another price (denoted v) per unit of steel. For the offer to be accepted, the price paid by the competitor must at least be equal to the turnover that the manufacturer could achieve by manufacturing automobiles itself. The competitor's problem can therefore be written as

$$\begin{aligned} &\min\ p = 400u + 600v \\ &\text{s.t.}\ \ u + v \geq 10{,}000 \\ &\qquad u + 2v \geq 16{,}000 \\ &\qquad u \geq 0,\ v \geq 0 \end{aligned} \qquad [1.24]$$

Graphical analysis finds the optimal solution $u = 4{,}000$ and $v = 6{,}000$, which corresponds to a global price of $p = 5{,}200{,}000$. Observe (and we will see later that this is no coincidence) that the optimal solution of the competitor's problem (the dual problem to the manufacturer's primal problem) coincides with the marginal prices of the manufacturer's problem, and the minimum price that the competitor can offer is equal to the manufacturer's maximum turnover.

1.12. Using Matlab

Matlab offers the *linprog* solver for optimizing LPs. The *linprog* solver takes the following input arguments:

– f: the objective function;

– A: the matrix of inequality constraints;

– b: the right-hand side of the inequality constraints;

– Aeq: the matrix of equality constraints;

– beq: the right-hand side of the equality constraints;

– lb: the vector of lower bounds of the decision variables;

– ub: the vector of upper bounds of the decision variables;

– x0: the starting point of x, only used for the active-set algorithm;

– options : options created with "optimoptions".

All *linprog* arguments use the following options, defined via the "optimoptions" input:

– Algorithm: the following algorithms can be used:

1) “interior-point” (default);

2) “dual-simplex”;

3) “active-set”;

4) “simplex”.

– LargeScale: the “interior-point” algorithm is used when this option is set to “on” (default) and the primal simplex algorithm is used when it is set to “off”.

– TolCon: the feasibility tolerance for constraints. The default value is 1e-4. It takes scalar values from 1e-10 to 1e-3 (only available for the dual simplex algorithm).

The output arguments of the *linprog* solver are as follows:

– fval: the value of the objective function at the solution x;

– x: the solution found by the solver. If exitflag > 0, then x is a solution; if not, x is the value of the variables when the solver terminated prematurely;

– exitflag: the reason why the algorithm terminated:

- 1: the solver converged to the solution x;
- 0: the number of iterations exceeded the “MaxIter option”;
- 2: no feasible point was found;
- 3: the LP problem is not bounded;
- 4: a NaN value was encountered when executing the algorithm;
- 5: the primal and dual problems are infeasible;
- 7: the search direction became too small and no other progress was observed.

The following list specifies the various syntaxes that can be used to call the *linprog* solver:

– *x = linprog(f, A, b)*: solves $\min f^T x$ such that Ax $\leq$ b (used when there are only equality constraints);

– *x = linprog(f, A, b, Aeq, beq)*: solves the problem with the equality constraints Aeqx = beq. Set A=[ ] and b=[ ] if no inequalities exist;

– *x = linprog(f, A, b, Aeq, beq, lb, ub)*: defines a set of lower and upper bounds on the design variables, x, so that the solution is always in the range $lb \leq x \leq ub$. Set Aeq=[ ] and beq=[ ] if no equalities exist;

– *x = linprog(f, A, b, Aeq, beq, lb, ub, x0)*: defines the starting point at x0 (only used for the active-set algorithm);

– *x = linprog(f, A, b, Aeq, beq, lb, ub, x0, options)*: minimizes with the specified options;

– *[x, fval] = linprog(...)*: returns the value of the objective function at the solution x;

– *[x, fval, exitflag] = linprog(...)*: returns a value exitflag that describes the exit condition;

– *[x, fval, exitflag, output] = linprog(...)*: returns a structure output that contains information about the optimization process;

– *[x, fval, exitflag, output, lambda] = linprog(...)*: returns a structure output whose fields contain the Lagrange multipliers λ at the solution x.

EXAMPLE 1.14.– Consider the linear program:

$$\begin{array}{lllr} \min & z = -2x_1 - 3x_2 + x_3 + 4x_4 + x_5 & & \\ \text{s.t.} & x_1 + 4x_2 - 2x_3 - x_4 & \leq & 8 \\ & x_1 + 3x_2 - 4x_3 - x_4 + x_5 & = & 10 \\ & -2x_1 - x_2 - 2x_3 - 3x_4 + x_5 & \leq & -20 \\ & x_1 & \geq & 5 \\ & x_2 & \geq & 1 \\ & x_j \geq 0;\ j = 1,2,3,4,5 & & \end{array} \qquad [1.25]$$

The following code can be executed to compute this example.

```
variables = {'x1', 'x2', 'x3', 'x4', 'x5'}; % construct the vector x
n = length(variables);
for i = 1:n % create the indices of x
eval([variables{i}, ' = ', num2str(i), ';']);
end
lb = zeros(1, n);
lb([x1, x2]) = [5, 1]; % add the bounds of the variables x1,x2
ub = Inf(1, n);
A = zeros(3, n); % create the matrix A
b = zeros(3, 1); % create the vector b
% define the constraints
A(1, [x1, x2, x3, x4]) = [1, 4, -2, -1]; b(1) = 8;
A(2, [x1, x2, x3, x4]) = [1, 3, 2, -1]; b(2) = 10;
A(3, [x1, x2, x3, x4, x5]) = [2, 1, 2, 3, -1]; b(3) = 20;
Aeq = zeros(1, n); % create the matrix Aeq
beq = zeros(1, 1); % create the vector beq
% define the equality constraints
Aeq(1, [x1, x2, x3, x4, x5]) = [1, 3, -4, -1, 1]; beq(1) = 7;
```

```
c = zeros(n, 1); % create the vector c
c([x1 x2 x3 x4 x5]) = [-2; -3; 1; 4; 1];
% call linprog solver
[x, objVal] = linprog(c, A, b, Aeq, beq, lb, ub);
for i = 1:n
fprintf('%s \t %20.4f\n', variables{i}, x(i))
end
fprintf(['The value of the objective function is' '%20.4f\n'],
        objVal)
```

For a color version of this code, see www.iste.co.uk/radi/optimizations.zip

Executing this code produces the following results:

```
Optimization terminated.
x1                 5.5000
x2                 1.0000
x3                 0.7500
x4                 0.0000
x5                 1.5000
The value of the objective function is           -11.7500
```

2

Integer Programming

2.1. Introduction

An integer linear program (ILP) is a program whose variables are restricted to integer values. In particular, the variables could simply be booleans, i.e. taking only the value 0 or 1, in which case the program is called a binary integer program (BIP). Many optimization problems require a solution with integer values. For example, in a production problem where we wish to find the optimal number of parts to manufacture, this number is often an integer for practical reasons.

Many constraints that seem nonlinear can be linearized because of integer variables. These possibilities greatly extend the field of applications of linear programming. Even if the resulting linear programs are often very difficult to solve, integer linear programming proves very useful in the context of modeling, enabling us to state discrete optimization problems concisely.

DEFINITION 2.1.– *A mixed integer program is a program where some (but not necessarily all) variables are restricted to integer values.*

The reader is welcome to refer to previous studies [SAL 75, NEM 98, WOL 98] for more details about integer programming.

2.2. Solving methods

Intuitively, it might seem tempting to consider the following immediate approach to solve an integer linear program:

– neglect the integrality constraints;

– round the solution.

Unfortunately, this approach generally does not work. The following example shows that the optimal integer is not always the closest integer to the optimal solution, if the optimal solution is a fraction.

EXAMPLE 2.1.– Maximize $10x_1 + 11x_2$ subject to $10x_1 + 12x_2 \leq 59$, $x_1, x_2 \geq 0$ with integer values.

The optimal solution x is given by $x = (x_1, x_2) = (5.9, 0)$. Indeed, $x_1 \leq 5.9 - 1.2x_2$, and we obtain the optimal solution by setting $x_1 = 5.9$, $x_2 = 0$. But the solution $x' = (6, 0)$ is not feasible, and the solution $x'' = (5, 0)$ is not optimal among the integer-valued solutions. The optimal integer-valued solution is $\tilde{x} = (1, 4)$.

The following remarks illustrate some of the difficulties of ILPs:

– the integer optimum is not necessarily a vertex of the polyhedron and can be inside of it;

– if the relaxed LP has an integer optimum, this integer is also the optimum of the ILP;

– if not, the optimal cost of the relaxed LP gives an upper bound (or lower bound for a minimization problem) for the optimal cost of the ILP;

– even if the polyhedron is non empty, it still might not admit any integer solutions;

– it is possible to construct examples where the optima of the ILP and the relaxed LP are arbitrarily far away from one another.

This chapter presents specialized methods for solving this type of program, including:

– the branch-and-bound method;

– the branch-and-cut method.

2.2.1. *Branch-and-bound method*

The branch-and-bound method essentially consists of the following three key elements:

1) a separation procedure (branching process): whenever we need to isolate a new subset of solutions, the branching process determines how many new subsets should be considered and how they should be defined;

2) an evaluation procedure (bounding process): a way to analyze the subproblems. The goal of this analysis is to evaluate the optimal value of the loss function of the subproblem; more precisely, we want to find a lower bound (or upper bound for

a maximization problem) for this value. The lower bound is typically obtained by relaxing the problem;

3) a pathing procedure: this indicates which subsets can be selected.

EXAMPLE 2.2.– Consider the LP (P):

$$\begin{aligned} \max\ & z(x_1, x_2) = x_1 + 4x_2 \\ \text{s.t.}\ & 5x_1 + 8x_2 \leq 40 \\ & -2x_1 + 3x_2 \leq 9 \\ & x_1, x_2 \geq 0 \text{ with integer values} \end{aligned} \qquad [2.1]$$

We proceed as follows:

1) solve (P) using the simplex algorithm while ignoring the constraint that x_1, x_2 must be integer valued. Optimal solution: $x = (x_1, x_2) = (1.55, 4.03)$, $z(x) = 17.67$. Branch into x_1: $x_1 \leq 1$ or $x_1 \geq 2$;

2) add the constraint $x_1 \leq 1$ to (P) and solve this relaxed problem using the simplex algorithm without the constraint that x_1, x_2 must be integer valued. Optimal solution: $x = (1, 3.67)$, $z(x) = 15.67$. Branch into x_2: $x_2 \leq 3$ or $x_2 \geq 4$;

3) add the constraints $x_1 \leq 1$ and $x_2 \leq 3$ to (P). The optimal solution of this program is integer valued: $x = (1, 3)$, $z(x) = 13$. Bound at 13 (i.e. only follow branches if the optimal solution is < 13);

4) add the constraints $x_1 \leq 1$ and $x_2 \geq 4$ to (P). This program has no feasible solutions;

5) add the constraint $x_1 \geq 2$ to (P). Optimal solution: $x = (2, 3.75)$, $z(x) = 17$. Branch into x_2: $x_2 \leq 3$ or $x_2 \geq 4$;

6) add the constraints $x_1 \geq 2$ and $x_2 \leq 3$ to (P). Optimal solution: $x = (3.2, 3)$, $z(x) = 15.2$. Branch into x_1: $x_1 \leq 3$ or $x_1 \geq 4$;

7) add the constraints $2 \leq x_1 \leq 3$ and $x_2 \leq 3$ to (P). Optimal solution: $x = (3, 3)$, $z(x) = 15$. Bound at 15;

8) add the constraints $x_1 \geq 4$ and $x_2 \leq 3$ to (P). This case is of no interest because $z(x) = 14 < 15$ for the optimal solution x;

9) add the constraints $x_1 \geq 2$ and $x_2 \geq 4$ to (P). No feasible solutions.

The optimal integer solution was therefore found in step (7): $x_1 = x_2 = 3$, with $z(x) = 15$.

REMARK.– The branch-and-bound method is not only used in linear programming but it can also be applied to any optimization problem involving a decision tree.

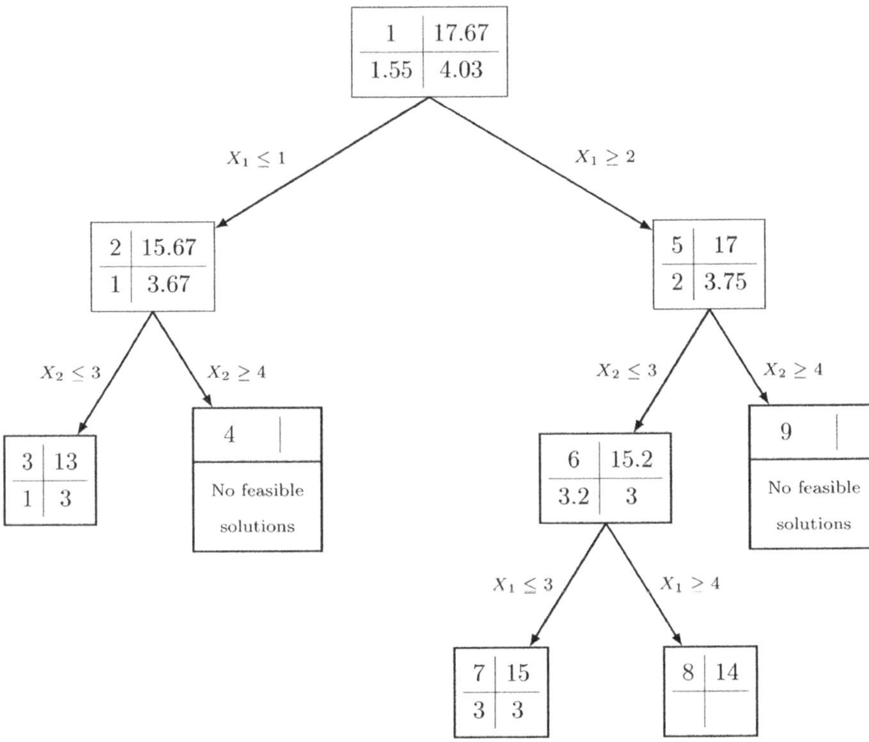

Figure 2.1. *Branch-and-bound method*

2.2.2. *The branch-and-cut method*

As above for the branch-and-bound method, let us illustrate this method with an example. Consider the linear program: maximize $z(x_1, x_2) = 2x_1 + x_2$ s.t. $x_1 - 4x_2 \leq 0$, $3x_1 + 4x_2 \leq 15$, $x_1, x_2 \geq 0$ with integer values.

The branch-and-cut method adds additional constraints that allow us to approach the integer-valued feasible solutions without eliminating them from the domain of feasible solutions. In the above example, one such constraint is given by $x_1 \leq 3$ (a cut is made at $x_1 = 3$).

How can we find these additional constraints if the linear program has more than two variables?

Consider the problem

$$(P)\begin{cases} \max\ z(x) = c^T x \\ \text{s.t.}\ \ Ax = b,\ x \geq 0 \end{cases}$$

where A is an $m \times n$ matrix.

Given a basis J (where $|J| = m$) and a feasible solution x, set

$$Ax = A^J x^J + A^{J^c} x^{J^c} = b, \text{ where } x^J = (x_j^J,\ j \in J),\ x^{J^c} = (x_j^{J^c},\ j \in J^c).$$

Since A^J is regular, $x^J + (A^J)^{-1} A^{J^c} x^{J^c} = (A^J)^{-1} b$. Set $\bar{A} = (A^J)^{-1} A^{J^c}$ and $\bar{b} = (A^J)^{-1} b$. Thus:

$$x^J + \bar{A} x^{J^c} = \bar{b}, \text{ where } x_i^J + \sum_{j \notin J} \bar{a}_{ij} x_j^{J^c} = \bar{b}_i \quad (i \in J),$$

$$\text{and } \bar{A} = (\bar{a}_{ij}, i, j = 1, \ldots, n).$$

Let $[\bar{a}_{ij}]$ be the integer part of $\bar{a}_{ij}$ and $< \bar{a}_{ij} >$ the fractional part, so that $\bar{a}_{ij} = [\bar{a}_{ij}] + < \bar{a}_{ij} >$.

EXAMPLE 2.3.– $[3.5] = 3$, $< 3.5 >= 0.5$ and $[-3.5] = -4$, $< -3.5 >= 0.5$.

It follows that:

$$x_i^J + \sum_{j \notin J} ([\bar{a}_{ij}] + < \bar{a}_{ij} >) x_j^{J^c} = [\bar{b}_i] + < \bar{b}_i > \qquad [2.2]$$

$$\Leftrightarrow \quad x_i^J + \sum_{j \notin J} [\bar{a}_{ij}] x_j^{J^c} - [\bar{b}_i] - \sum_{j \notin J} < \bar{a}_{ij} x_j^{J^c}.$$

Let x be a feasible solution of (P) with integer values. The equality [2.3] holds for $\underline{x}$, as the left-hand side is integer valued. Furthermore,

$$\underline{x}_i^J + \sum_{j \notin J} [\bar{a}_{ij}] \underline{x}_j^{J^c} \leq \underline{x}_i^J + \sum_{j \notin J} \bar{a}_{ij} \underline{x}_j^{J^c} = \bar{b}_i. \qquad [2.3]$$

Since the left part of [2.3] is integer valued,

$$\underline{x}_i^J + \sum_{j \notin J} [\bar{a}_{ij}] \underline{x}_j^{J^c} \leq [\bar{b}_i].$$

Since [2.3] holds for every feasible solution, it follows that:

$$< \bar{b} > - \sum_{j \notin J} < \bar{a}_{ij} > \underline{x}_j^{J^c} \leq 0.$$

THEOREM 2.1.– Adding the constraint

$$\sum_{i \notin J} < \bar{a}_{ij} > x_i^{J^c} \geq < \bar{b}_i >, \quad (i \in J)$$

to (P) does not eliminate any feasible integer-valued solutions from the set of feasible solutions.

Therefore, to solve (P), we can proceed as follows:

1) Solve with the simplex algorithm without the condition, "x_1, x_2 have integer values". Optimal solution is: $x = (\frac{18}{7}, \frac{18}{7}, 0, 0)^T$.

Initial tableau

x_1	x_2	x_3	x_4	
-1	1	1	0	$x_1^J = x_3 = 0$
5	2	0	1	$x_2^J = x_4 = 18$
-2	-1	0	0	$z(x) = 0$

Final tableau

x_1	x_2	x_3	x_4	
0	1	$\frac{5}{7}$	$\frac{1}{7}$	$x_1^J = x_2 = \frac{18}{7}$
1	0	$-\frac{2}{7}$	$\frac{1}{7}$	$x_2^J = x_1 = \frac{18}{7}$
0	0	$\frac{1}{7}$	$\frac{3}{7}$	$z(x) = \frac{54}{7}$

The new constraint is given as:

$$\frac{5}{7}x_3 + \frac{1}{7}x_4 \geq \frac{4}{7} \quad (S1)$$

To express (S1) in terms of the initial variables x_1, x_2, observe that

$$\frac{18}{7} = x_2 + \frac{5}{7}x_3 + \frac{1}{7}x_4 \geq x_2 + \frac{4}{7} \Rightarrow x_2 \leq 2.$$

2) Solve (P) together with (S1) by the simplex algorithm. We need to solve an auxiliary program first.

Auxiliary program, initial tableau

c	0	0	$\frac{5}{7}$	$\frac{1}{7}$	-1	0
Basic variables	x_1	x_2	x_3	x_4	x_5	y
$x_1^J = x_2 = \frac{18}{7}$	0	1	$\frac{5}{7}$	$\frac{1}{7}$	0	0
$x_2^J = x_1 = \frac{18}{7}$	1	0	$-\frac{2}{7}$	$\frac{1}{7}$	0	0
$x_3^J = y = \frac{4}{7}$	0	0	$\frac{5}{7}$	$\frac{1}{7}$	-1	1
$z''(x) = -\frac{4}{7}$	0	0	$-\frac{5}{7}$	$-\frac{1}{7}$	1	0

Auxiliary program, final tableau

c	0	0	$\frac{5}{7}$	$\frac{1}{7}$	-1
Basic variables	x_1	x_2	x_3	x_4	x_5
$x_1^J = x_2 = 2$	0	1	0	0	1
$x_2^J = x_1 = \frac{14}{5}$	1	0	0	$\frac{1}{5}$	$-\frac{2}{5}$
$x_3^J = x_3 = \frac{4}{5}$	0	0	1	$\frac{1}{5}$	$-\frac{7}{5}$
$z''(x) = 0$	0	0	0	0	0

Since $z''(x) = 0$, the program (P)+(S1) has a feasible solution. This solution is namely $x^J = (x_1, x_2, x_3)^T = (\frac{14}{5}, 2, \frac{4}{5})^T$.

Final tableau, second phase

0 0 0 $-\frac{2}{5}$ $-\frac{1}{5}$	c
x_1 x_2 x_3 x_4 x_5	Basic variables
0 1 0 0 1	$x_1^J = x_2 = 2$
1 0 0 $\frac{1}{5}$ $-\frac{2}{5}$	$x_2^J = x_1 = \frac{14}{5}$
0 0 1 $\frac{1}{5}$ $-\frac{7}{5}$	$x_3^J = x_3 = \frac{4}{5}$
0 0 0 $\frac{2}{5}$ $\frac{1}{5}$	$z = \frac{38}{5}$

The optimal solution is equal to $x = (\frac{14}{5}, 2, \frac{4}{5}, 0, 0)^T$. The new constraint is:

$$\frac{1}{5}x_4 + \frac{3}{5}x_5 \geq \frac{4}{5} \quad (S2)$$

Hence: $x_1 + x_2 \leq 4$.

3) Solve (P) together with (S1) and (S2). This requires another auxiliary program.

Final tableau, second phase of the simplex algorithm

0 0 0 $-\frac{1}{3}$ 0 $-\frac{1}{3}$	c
x_1 x_2 x_3 x_4 x_5 x_6	Basic variables
0 1 0 $-\frac{1}{3}$ 0 $\frac{5}{3}$	$x_1^J = x_2 = \frac{2}{3}$
1 0 0 $\frac{1}{3}$ 0 $-\frac{2}{3}$	$x_2^J = x_1 = \frac{10}{3}$
0 0 1 $\frac{2}{3}$ 0 $-\frac{7}{3}$	$x_3^J = x_3 = \frac{8}{3}$
0 0 0 $\frac{1}{3}$ 1 $-\frac{5}{3}$	$x_4^J = x_5 = \frac{4}{3}$
0 0 0 $\frac{1}{3}$ 0 $\frac{1}{3}$	$z = \frac{22}{3}$

The optimal solution is equal to $x = (\frac{10}{3}, \frac{2}{3}, \frac{8}{3}, 0, \frac{4}{3}, 0)^T$. The new constraint is:

$$x_4 + x_6 \geq 1 \quad (S3)$$

Expressed in terms of x_1, x_2, (S3) becomes: $2x_1 + x_2 \leq 7$.

4) Solve (P) together with (S1), (S2) and (S3). This requires another auxiliary program. This time, it gives us the integer-valued optimal solution $x^* = (3, 1, 2, 0, 1, 1)^T$. The optimal integer-valued solution of (P) is therefore given by $x_1 = 3$, $x_2 = 1$ and $z^*(x, y) = 7$.

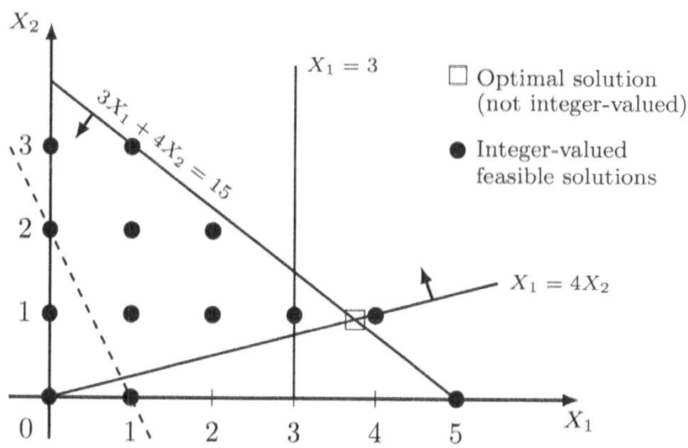

Figure 2.2. *Branch-and-cut method*

2.3. Binary programming

The knapsack problem is one of the most common problems in binary programming.

2.3.1. *Knapsack problem*

John the camper embarks on a hike in the mountains. His knapsack can only carry a limited weight P. Each item i that he could potentially bring with him weighs p_i and provides a utility u_i for John's hike. Which items should John bring with him to maximize the utility without exceeding the weight limit?

The decision variables are given as:

– $x_i = 1$ if John brings item i with him;

– $x_i = 0$ otherwise.

The problem can be stated in the following form:

$$\max \quad u_1x_1 + u_2x_2 + \cdots + u_nx_n \qquad [2.4]$$

$$\text{s.t. } p_1x_1 + p_2x_2 + \cdots + p_nx_n \leq P \qquad [2.5]$$

$$x_1, \ldots, x_n \in \{0, 1\} \qquad [2.6]$$

REMARK.– In general, this type of problem is known as the knapsack problem. There are 2^n ways to choose a subset of n elements. If we assume that it takes $\frac{1}{100}$ seconds to evaluate one subset, it would take more than an entire year to evaluate every combination if there were just 32 possible elements. Considering every possibility exhaustively is therefore impossible. To be efficient, an algorithm needs to construct an optimal solution by examining a very small number of solutions.

2.3.2. *Investment problem*

A company has 1,400,000 euros that it wishes to invest. The experts propose four possible investments as follows:

	Cost	Return	Yield
Inv. 1	500,000	1,600,000	3.20
Inv. 2	700,000	2,200,000	3.14
Inv. 3	400,000	1,200,000	3.00
Inv. 4	300,000	800,000	2.67

The decision variables are x_i, $i = 1, \ldots, 4$:

– $x_i = 1$ if investment i is chosen;

– $x_i = 0$ otherwise.

The goal is to maximize the return:

$$\max 16x_1 + 22x_2 + 12x_3 + 8x_4$$

which is subject to the constraint of the investment budget:

$$5x_1 + 7x_2 + 4x_3 + 3x_4 \leq 14.$$

This leads to the following binary program:

$$\max\ 16x_1 + 22x_2 + 12x_3 + 8x_4$$

$$\text{s.t.}\ \ 5x_1 + 7x_2 + 4x_3 + 3x_4 \leq 14 \qquad [2.7]$$

$$x_1, x_2, x_3, x_4 \in \{0, 1\}$$

The solution is $x_1 = 0$, $x_2 = 1$, $x_3 = 1$, $x_4 = 1$ with a return of 4,200,000 euros and a cost of 1,400,000 euros.

To find the solution, we can use linear relaxation, which is solved efficiently by a greedy algorithm: simply take the items with the best return until the capacity is reached.

	Cost	Return	Yield
Inv. 1	500,000	1,600,000	3.20
Inv. 2	700,000	2,200,000	3.14
Inv. 3	400,000	1,200,000	3.00
Inv. 4	300,000	800,000	2.67

Relaxation of $F(U = -\infty)$:

		Cost	Return	Yield
1	Inv. 1	500,000	1,600,000	3.20
1	Inv. 2	700,000	2,200,000	3.14
0.5	Inv. 3	400,000	1,200,000	3.00
	Inv. 4	300,000	800,000	2.67

Relaxation of F: $x^* = (1, 1, 0.5, 0)^T$, we have: $b(F) = 4\,400\,000 > U$ (because maximization!) and $F_1: x_3 = 1 \quad F_2: x_3 = 0$.

Relaxation of $F_1 (U = -\infty)$:

		Cost	Return	Yield
1	Inv. 1	500,000	1,600,000	3.20
1	Inv. 2	700,000	2,200,000	3.14
$\frac{5}{7}$	Inv. 3	400,000	1,200,000	3.00
0	Inv. 4	300,000	800,000	2.67

Relaxation of F_1: $x^* = (1, 5/7, 1, 0)$, we have: $b(F_1) = 4371429 > U$ and $F_{11}: x_2 = 0$ $F_{12}: x_2 = 1$.

Relaxation of $F_{11} (U = -\infty)$:

		Cost	Return	Yield
1	Inv. 1	500,000	1,600,000	3.20
1	Inv. 2	700,000	2,200,000	3.14
0	Inv. 3	400,000	1,200,000	3.00
0	Inv. 4	300,000	800,000	2.67

Relaxation of F_{11}: $x^* = (1, 0, 1, 1)$, we have: $b(F_{11}) = 3,600,000 > U$ with $U = 3,600,000$.

Relaxation of F_{12} $(U11 = 3,600,000)$:

		Cost	Return	Yield
$\frac{3}{5}$	Inv. 1	500,000	1,600,000	3.20
1	Inv. 2	700,000	2,200,000	3.14
0	Inv. 3	400,000	1,200,000	3.00
0	Inv. 4	300,000	800,000	2.67

Relaxation of F_{12}: $x^* = (3/5, 1, 1, 0)$:

$b(F_{12}) = 4,360,000 > U$

$F_{121}: x_1 = 0$ $F_{122}: x_1 = 1$

Relaxation of $F_{121}(U_{11} = 3,600,000)$:

		Cost	Return	Yield
$\frac{3}{5}$	Inv. 1	500,000	1,600,000	3.20
1	Inv. 2	700,000	2,200,000	3.14
0	Inv. 3	400,000	1,200,000	3.00
0	Inv. 4	300,000	800,000	2.67

Relaxation of F_{121}: $x^* = (0, 1, 1, 1)^T$, we have: $b(F_{121}) = 4,200,000 > U$ with $U = 4,200,000$.

Relaxation of $F_{122}(U_{121} = 4,200,000)$:

		Cost	Return	Yield
1	Inv. 1	500,000	1,600,000	3.20
1	Inv. 2	700,000	2,200,000	3.14
1	Inv. 3	400,000	1,200,000	3.00
?	Inv. 4	300,000	800,000	2.67

Relaxation of F_{122}: inadmissible.

Deletion of F_{122} and relaxation of $F_2(U_{121} = 4,200,000)$:

		Cost	Return	Yield
1	Inv. 1	500,000	1,600,000	3.20
1	Inv. 2	700,000	2,200,000	3.14
1	Inv. 3	400,000	1,200,000	3.00
$\frac{2}{3}$	Inv. 4	300,000	800,000	2.67

Relaxation of F_2: $x^* = (1, 1, 0, 2/3)$, we have: $b(F_2) = 4,333,333 > U$,

$F_{21} : x_4 = 0$ $F_{22} : x_4 = 1$ then relaxation of $F_{21}(U_{121} = 4,200,000)$:

		Cost	Return	Yield
1	Inv. 1	500,000	1,600,000	3.20
1	Inv. 2	700,000	2,200,000	3.14
0	Inv. 3	400,000	1,200,000	3.00
0	Inv. 4	300,000	800,000	2.67

Relaxation of F_{21}: $x^* = (1,1,0,0)^T$, we have: $b(F_{21}) = 3,800,000 \leq U$.

Deletion of F_{21} and relaxation of $F_{22}(U_{121} = 4,200,000)$:

		Cost	Return	Yield
1	Inv. 1	500,000	1,600,000	3.20
0	Inv. 2	700,000	2,200,000	3.14
$\frac{6}{7}$	Inv. 3	400,000	1,200,000	3.00
0	Inv. 4	300,000	800,000	2.67

Relaxation of F^{22}: $x^* = (1,6/7,0,1)$, we have: $b(F_{22}) = 4,285,714 > U$.

$F_{221}: x_2 = 0$ $F_{222}: x_2 = 1$ then relaxation of $F_{221}(U_{121} = 4,200,000)$:

		Cost	Return	Yield
1	Inv. 1	500,000	1,600,000	3.20
0	Inv. 2	700,000	2,200,000	3.14
0	Inv. 3	400,000	1,200,000	3.00
1	Inv. 4	300,000	800 000	2.67

Relaxation of F_{221}: $x^* = (1,0,0,1)$, we have: $b(F_{221}) = 2,400,000 \leq U$.

Deletion of F_{221} and relaxation of $F_{222}(U_{121} = 4,200,000)$:

		Cost	Return	Yield
$\frac{4}{5}$	Inv. 1	500,000	1,600,000	3.20
0	Inv. 2	700,000	2,200,000	3.14
0	Inv. 3	400,000	1,200,000	3.00
1	Inv. 4	300,000	800,000	2.67

Relaxation of F_{222}: $x^* = (4/5, 1, 0, 1)$, we have: $b(F_{222}) = 4,280,000 > U$ and

$F_{2221}: x_1 = 0 \qquad F_{2222}: x_1 = 1.$

Relaxation of $F_{222}(U_{121} = 4,200,000)$:

		Cost	Return	Yield
?	Inv. 1	500,000	1,600,000	3.20
1	Inv. 2	700,000	2,200,000	3.14
0	Inv. 3	400,000	1,200,000	3.00
1	Inv. 4	300,000	800,000	2.67

Relaxation of $F_{2221}: x^* = (0, 1, 0, 1)^T$, we have: $b(F_{2221}) = 3\,000\,000 \leq U$ and F_{2222} inadmissible.

REMARK.–

1) Only 7 combinations were considered (F_{11}, F_{121}, F_{122}, F_{21}, F_{221}, F_{2221} and F_{2222}). A full enumeration would have required us to consider $2^4 = 64$ combinations.

2) In general, the greedy algorithm works by gradually building up a solution by making choices that seem optimal locally. In some cases, this yields the best solution; this is known as an exact greedy algorithm. In other cases, it does not; this is described as greedy heuristics.

2.4. Decomposition principle

Consider an ILP (P) defined with integer (or mixed) variables $x \in \mathbb{N}^n$ (or $x \in \mathbb{R}^n$):

$$(P) \begin{cases} \min & c^T x \\ \text{s.t.} & Ax \geq b \\ & x \in \mathbb{N}^n \end{cases}$$

In practical problems, the coefficient matrix A often has a block diagonal structure, where the gray zones represent non-zero coefficients (see Figure 2.3). Only the bottom block involves every variable of the problem: the constraints in this block are called the coupling constraints. The other diagonal blocks are smaller and only involve a limited set of variables that appear exclusively in their respective blocks.

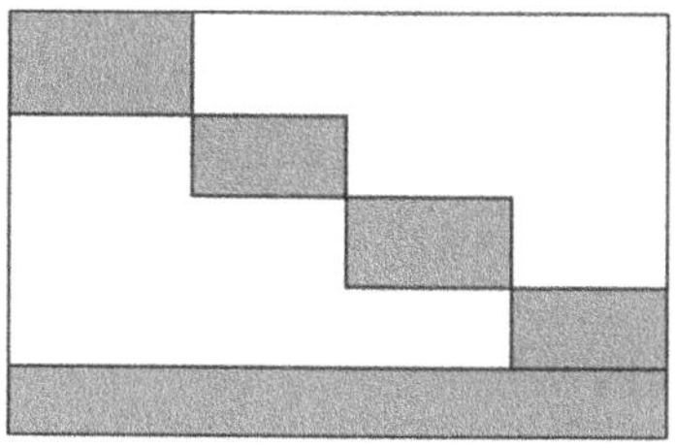

Figure 2.3. *Matrix A*

DEFINITION 2.2.– *Solving an ILP without taking advantage of this matrix composition is known as frontal solving.*

DEFINITION 2.3.– *Using the block structure of an ILP to solve it by dividing the solving procedure into multiple programs and subprograms is known as decomposition.*

The master problem of the ILP is obtained from the original ILP and gives the solution to the original problem. It generally keeps the same coupling constraints and plays a piloting role for the decomposition method and the auxiliary (or satellite) problems considered in subprograms to solve the problem in full detail. Typically, each subproblem is associated with an isolated block in the coefficient matrix M.

The master program and the satellite programs are then incorporated together into an algorithmic process that converges to a lower bound or the optimal solution of the problem. Depending on how the problem is decomposed and the choice of mathematical technique, various decompositions are possible, and a variety of algorithms have been developed accordingly.

Some key examples include:

– decomposition by Lagrangian relaxation, also known as decomposition by price: Lagrangian relaxation is applied to the coupling constraints by adding them to the objective function. This gives a master problem that pilots the solving procedure by fixing a price (the Lagrangian cost) associated with the coupling inequalities. This master problem is a set of independent ILPs: each ILP is viewed as a subproblem. Note that this method only converges to a lower bound of the problem and not to a solution;

– column generation and branching reformulates problems with an exponential number of variables to construct a problem with a decomposable matrix. The column generation algorithm can be understood as a decomposition algorithm where the master problem pilots the subproblems via the dual costs of its inequalities;

– Dantzig–Wolfe decomposition: there is a quasi-automatic technique to reformulate an ILP as a decomposable program with an exponential number of variables. The resulting decomposition, called a Dantzig–Wolfe decomposition, is then solved by column generation and branching (the two terms are often used synonymously);

– Benders decomposition [BEN 62]: there is another quasi-automatic technique to reformulate an ILP as a decomposable program with an exponential number of inequalities. The resulting decomposition, called a Benders decomposition, is then solved by a method of (possibly non-polynomial) cuts and branching.

A dedicated study is required each time a decomposition is implemented. The decomposition techniques listed above have found great success with very large scale industrial problems.

2.4.1. *Benders decomposition*

This method is dedicated to mixed problems. It proceeds by solving a master problem containing only real variables and multiple subproblems with integer variables. At each iteration, the master problem gives a lower bound of the optimal solution, and each subproblem gives an upper bound. The method can be justified as follows, distinguishing the real and integer variables:

$$\begin{array}{ll} \max_{x,y} & c_1x + c_2y \\ C1: & A_1x + A_2y \leq b \\ C2: & x \geq 0 \\ C3: & y \geq 0, \text{ integers} \end{array} \qquad [2.8]$$

To find a lower bound, note that we can rewrite the above linear program by fixing y. The LP then becomes:

$$\begin{array}{ll} \max_{x} & c_1x + c_2y \\ C1: & A_1x \leq b - A_2y \\ C2: & x \geq 0 \end{array} \qquad [2.9]$$

Write $V^0(y)$ for the optimal solution. Let (D1) be the dual of this program:

$$(D1)\begin{cases} \min_u \ u(b - A_2y) \\ C1: \ u.A_1 \geq c_1 \\ C2: \ u \geq 0 \end{cases} \qquad [2.10]$$

Write $W^0(y)$ for its optimal solution, which clearly satisfies $W^0(y) + c_2y = V^0(y)$. Therefore, since y is fixed, $W^0(y) + c_2y = V^0(y)$ is a lower bound of the desired solution.

To obtain an upper bound, observe that it would be possible to solve the initial program by enumerating each of the points y. This program can therefore be rewritten as follows:

$$\begin{array}{l} \max \ Z \\ C1: \ Z \leq c_2.y + u_1(b - A_1x) \\ C2: \ Z \leq c_2.y + u_2(b - A_2x) \\ \ .: \ \ \ldots \\ Ck: \ Z \leq c_2.y + u_k(b - A_kx) \\ C5: \ y \geq 0, \ (integers) \end{array} \qquad [2.11]$$

Since it is impossible to generate all of these constraints, we will generate them successively and hope to find the optimal solution after only a few of them have been added. Write P^r for the linear program restricted to r constraints:

$$(P^r)\begin{cases} \max \ Z \\ C1: \ Z \leq c_2.y + u_1(b - A_1x) \\ C2: \ Z \leq c_2.y + u_2(b - A_2x) \\ \ .: \ \ \ldots \\ Ck: \ Z \leq c_2.y + u_r(b - A_rx) \\ C5: \ y \geq 0, \ (integers) \end{cases} \qquad [2.12]$$

Solving this problem clearly gives an upper bound for the original problem, since only r constraints are present among the k necessary constraints. Note that, in the case of a minimization problem, the roles of the master problem and subproblems are switched: the master problem gives an upper bound for the starting problem.

Consider the following LP (P) proposed by Jensen [JEN n/a]:

$$(P)\begin{cases} \max \ 8x_1 + 6x_2 - 2x_3 - 42y_1 - 18y_2 - 33y_3 \\ C1: \ 2x_1 + x_2 - x_3 - 10y_1 - 8y_2 \leq -4 \\ C2: \ x_1 + x_2 + x_3 - 5y_1 - 8y_3 \leq -3 \\ C3: \ x_i \geq 0 \\ C3: \ y_i \geq 0, \ y_i \leq 1, \ \text{integers} \end{cases} \qquad [2.13]$$

This LP admits the optimal solution $x^* = (0, 4.5, 0.5)^T$ and $y^* = (0, 1, 1)^T$ with cost function $z^* = -25$. The matrices used in the LP (P) are as follows:

$$c_1 = \begin{pmatrix} 8 & 6 & -2 \end{pmatrix}^T \quad c_2 = \begin{pmatrix} -42 & -18 & -33 \end{pmatrix}^T$$

$$A_1 = \begin{pmatrix} 2 & 1 & -1 \\ 1 & 1 & 1 \end{pmatrix} \quad A_2 = \begin{pmatrix} -10 & -8 & 0 \\ -5 & 0 & -3 \end{pmatrix} \quad b = \begin{pmatrix} -4 \\ -3 \end{pmatrix}$$

The model (D1) has two variables: $u = (u_1, u_2)^T$. The coefficients of the objective function are given as:

$$b - A_2 x = \begin{pmatrix} -4 + 10y_1 + 8y_2 \\ -3 + 5y_1 + 8y_3 \end{pmatrix}.$$

The objective function of (D1) is therefore:

$$W^0(y) = \min u_1(-4 + 10y_1 + 8y_2) + u_2(-3 + 5y_1 + 8y_3).$$

The constraints of (D1) are given as:

$$\begin{array}{ll} (1) & 2u_1 + u_2 \geq 8 \\ (2) & u_1 + u_2 \geq 6 \\ (3) & -u_1 + u_2 \geq -2 \\ (4) & u_1 + u_2 \leq 9999 \\ & u_1 \geq 0, \ u_2 \geq 0. \end{array}$$

The constraint (4) is added to guarantee that the solution is bounded.

The constraints of the integer program are formulated from the solutions of (D1). The general form of the constraint is

$$\begin{aligned} Z &\leq c_2 y + u_k(b - A_2 y), \\ Z &\leq (c_2 - u_k A_2)y + u_k b \\ (-c_2 + u_k A_2)y + Z &\leq u_k b \end{aligned}$$

where u_k is a solution of (D1). Substituting the relevant vectors and matrices gives:

$$Z \leq -42y_1 - 18y_2 - 33y_3 + u_1(-4 + 10y_1 + 8y_2) + u_2(-3 + 5y_1 + 8y_3).$$

Combining terms and moving any terms with variables to the left-hand side gives:

$$y_1(42 - 10u_1 - 5u_2) + y_2(18 - 8u_1) + y_3(33 - 8u_2) + Z \leq -4u_1 - 3u_2.$$

The various steps of the approach are as follows:

– Step 0: solve (D1) with an objective function of zero; the solution is $u = (4, 2)$, so we have a new constraint

$$-8y_1 - 14y_2 + 17y_3 + Z \leq -22.$$

– Step 1: solve (P[1]) with only this constraint, giving the solution $y = (1,1,0)^T$ and $Z = 0$.

– Step 2: solve (D1) with the new y :

$$W^0 = +14u_1 + 2u_2.$$

The optimal solution is $u = (0,8)$, with a cost of 16.

– Step 3: optimality test: $-44 \leq Z^* \leq 0$. Construct a new constraint with the new values of u:

$$2y_1 + 18y_2 - 31y_3 + Z \leq -24.$$

Begin a new sequence of steps with the problem (P^2):

– Step 1: solve the problem (P^2) with both constraints. The optimal solution is $y = (1,1,1)^T$ and $Z = -17$.

– Step 2: construct the new objective function of (D1) with the new y:

$$W^0 = +14u_1 + 10u_2,$$

then solve the problem (D1). The optimal solution is $u(2,4)$ with $Z = 68$.

– Step 3: optimality test $-25 \leq Z^* \leq -17$. Construct a new constraint with the new u:

$$2y_1 + 2y_2 + y_3 + Z \leq -20.$$

Next, construct the new problem (P^3) with the three new constraints:

– Step 1: solve (P^3); we find $y = (0,0,0)^T$ with $Z = -24$.

– Step 2: construct the new objective function of (D1) with the new y:

$$W^0 = -4u_1 - 3u_2,$$

then solve the problem (D1). We find $u = (500.5, 4998.5)$ with $W^0 = -34997.5$.

– Step 3: optimality test $-25 \leq Z^* \leq -24$. Construct the new constraint:

$$-74955.5y_1 - 39986y_2 - 3995517y_3 + Z \leq -34997.5.$$

Now construct the problem (P^4) with the four constraints on y.

– Step 1: solve the problem (P^4). We find $y = (1,1,1)^T$ and $Z = -25$.

– Step 2: construct the new objective function of the problem (D1) with the new y:

$$W^0 = 14u_1 + 10u_2,$$

then solve the problem (D1). We find $u = (2,4)$ with $W^0 = 68$.

– Step 3: optimality test $-25 \leq Z^* \leq -25$. We have found the optimal solution of the starting problem with the integer-valued solution $y = (1,1,1)^T$ with $Z^* = -25$. The actual solution is $x^* = (4,6,0)^T$.

2.5. Using *Matlab*

To work with integer-valued and mixed programs, Matlab offers the *intlinprog* function, which has the following syntax:

– x = *intlinprog(f,intcon,A,b)* solves $\min f' * x$ such that the components of x in intcon are integers and $A * x \leq b$;

– x = *intlinprog(f,intcon,A,b,Aeq,beq)* solves the problem above while additionally satisfying the equality constraints $Aeq * x = beq$. Set $A = []$ and $b = []$ if no inequalities exist;

– x = *intlinprog(f,intcon,A,b,Aeq,beq,lb,ub)* defines a set of lower and upper bounds on the design variables, x, so that the solution is always in the range $lb \leq x \leq ub$. Set $Aeq = []$ and $beq = []$ if there are no equality constraints;

– x = *intlinprog(f,intcon,A,b,Aeq,beq,lb,ub,x0)* optimizes using an initial feasible point x_0. Set $lb = []$ and $ub = []$ if there are no bounds;

– x = *intlinprog(f,intcon,A,b,Aeq,beq,lb,ub,x0,options)* minimizes using the optimization options specified in options. Use optimoptions to set these options. Set $x_0 = []$ if no initial point exists;

– x = *intlinprog(problem)* uses a problem structure to encapsulate all solver inputs. You can import a problem structure from an MPS file using *mpsread*. You can also create a problem structure from an OptimizationProblem object by using prob2struct;

– *[x,fval,exitflag,output] = intlinprog(...)*, for any input arguments described above, returns fval = $f' * x$, a value exitflag describing the exit condition, and a structure output containing information about the optimization process.

EXAMPLE 2.4.– Consider the mixed LP:

$$\begin{aligned} \min\ & 8x_1 + x_2 \\ \text{s.t. } & x_1 + 2x_2 \geq -14 \\ & -4x_1 - x_2 \leq -33 \\ & 2x_1 + x_2 \leq 20 \\ & x_2 \text{ integer} \end{aligned} \qquad [2.14]$$

```
>> f = [8;1];
intcon = 2;
>> A = [-1,-2;
```

```
    -4,-1;
    2,1];
b = [14;-33;20];
```

Executing the function *intlinprog* gives:

```
>> x = intlinprog(f,intcon,A,b)
LP:                Optimal objective value is 59.000000.
Optimal solution found.

Intlinprog stopped at the root node because the objective value
is within a gap tolerance of the optimal value,
options.TolGapAbs = 0 (the default value). The intcon variables
are integer within tolerance, options.TolInteger = 1e-05
(the default value).


x =

    6.5000
    7.0000
```

The optimal solution is $x^* = (6.5, 7)^T$ with an optimal cost of $z^* = 59$.

EXAMPLE 2.5.– Consider the mixed problem:

$$
\begin{array}{ll}
\min & -3x_1 - 2x_2 - x3 \\
\text{s.t.} & x_1 + x_2 + x_3 \leq 7 \\
 & 4x_1 + 2x_2 + x_3 = 2 \\
 & x_1, x_2 \geq 0 \\
 & x_3 \text{ binary}
\end{array}
\qquad [2.15]
$$

```
>> f = [-3;-2;-1];
intcon = 3;
>> A = [1,1,1];
b = 7;
>> Aeq = [4,2,1];
beq = 12;
>> lb = zeros(3,1);
ub = [Inf;Inf;1]; % forces x(3) to be binary
```

After executing the function *intlinprog*, we find:

```
>> x = intlinprog(f,intcon,A,b,Aeq,beq,lb,ub)
LP:                Optimal objective value is -12.000000.


Optimal solution found.

Intlinprog stopped at the root node because the objective value '
is within a gap tolerance of the optimal value,
options.TolGapAbs = 0 (the default value). The intcon variables
are integer within tolerance, options.TolInteger = 1e-05
(the default value).


x =

         0
    5.5000
    1.0000
```

The optimal solution is therefore $x^* = (0, 5.5, 1)^T$ with an optimal cost of $z^* = -12$.

3

Dynamic Programming

3.1. Introduction

Sometimes, when attempting to solve a problem of size n by the top-down approach (divide and solve), the decomposition generates multiple identical subproblems. If we solve each subexample separately, without taking advantage of this redundancy, the resulting algorithm is very inefficient. However, if we are careful to only solve each distinct subexample once (for example by saving any results calculated earlier), the algorithm can be made efficient. This is the general idea of dynamic programming (DP). The method works because of Bellman's principle of optimality [BEL 56]: the optimal solution for a problem contains the optimal solutions for all subproblems. Thus, DP solves optimization problems that satisfy Bellman's optimality principle by taking advantage of a recursive formula.

In other words, DP is a mathematical technique whose goal is to make sequential decisions independently of one another. It can be used to solve any discrete optimization problem [SNI 06] by implicitly enumerating the solutions. Although it was already being used at the time, it was popularized as a general method by Bellman, who formalized and named the approach. Since then, it has constantly evolved, and extensive research has focused on analyzing it as a solving method, and using it to solve optimization problems. DP has allowed elegant solutions to be found for a range of problems across a variety of fields (operations research, learning, bioinformatics, system control, linguistics, game theory, Markov processes, etc.). It immediately found success in practice, since many of the loss functions used by the industry were of this form, for example when handling and optimizing chemical processes, or managing inventory [COR 02].

The method has several applications. A non-exhaustive list of examples might include:

– the shortest path problem (Bellman–Ford algorithm and Floyd–Warshall);

– the resource allocation problem. Suppose (for example) that we wish to allocate m skis to n skiers ($m > n$) while minimizing the size difference between skiers and skis. This problem satisfies the substructure optimality property (if a distribution is optimal, then any subset of skis and skiers is optimal), which means that it can be solved by DP;

– the knapsack problem. This is a classical problem in operations research that is NP-hard but which can be solved pseudo-polynomially with a DP algorithm;

– finding the longest strictly increasing subsequence of a sequence of numbers;

– text algorithm problems, such as calculating the longest subsequence shared by two chains, calculating the Levenshtein distance, or sequence alignment (for example with the Smith–Waterman algorithm);

– in economics: models of growth, inventory management, natural resource exploitation and the environment [LUC 89].

DP can be applied to problems that satisfy the two following properties:

– the optimality property: in an optimal sequence (of decisions or choices), each subsequence must also be optimal;

– the overlapping subproblem property: the subproblems are not disjoint, meaning that they have shared subproblems [LEI 01].

3.2. Solving strategy

The strategy of DP can be summarized into three key points:

1) find the solution as a recurrence relation or recursive algorithm;

2) justify why the number of possible cases at each step is bounded by a polynomial;

3) specify an order of evaluation for intermediate results so that any required results are available whenever needed.

The goal is to subdivide the problem into subproblems, then organize how these subproblems are solved in such a way that each of them is evaluated only once. More precisely, we are looking for an optimal solution given a certain cost. This can be done in two ways: top-down or bottom-up.

– *Top-down method*: also known as the memorization method or recursive method. This method handles the recursive tree top-down by following a path of postfixed depth. To avoid redundancy in the computation, the algorithm is parameterized in such a way that it remembers the solutions of any previously solved subproblems.

– *Bottom-up method*: the idea of this approach is to choose a favorable order for solving the subproblems. This order must guarantee that each subproblem is solved only once, and that it is only computed after computing all of the subproblems on which it depends. To do this, the problem is divided into steps. Each step consists of several subproblems, associated with a solving strategy and a certain number of states. The solution of each subproblem of a given step only depends on the subproblems of previous steps. The subproblems are solved bottom-up, from the first step to the last.

The relationships between subproblems can be represented as a multilevel direct acyclic graph (DAG). It is structured as follows:

– each step (or level) is a set of nodes;

– each node represents a set of subproblems. The node with no outgoing edges represents the original problem being solved, and the nodes with no incident edges are the initial elements of the problem;

– if there is an edge from a node N_1 to a node N_2, then computing the optimal solutions of the subproblems corresponding to node N_2 is dependent on computing the subproblems corresponding to node N_1;

– any two nodes that are not connected by an edge are said to be independent nodes.

A level is a set of mutually independent subproblems. If a subproblem belongs to a step i, the node that represents it does not have any edges to the nodes of any step j ($j \leq i$) and contains at least one edge to a node of step $(i + 1)$. The resolution of a DP problem is reduced to the calculation of DAG-multi-level node values. In the literature, this graph is called the "task graph" or "dynamic programming graph".

Any algorithm that solves the subproblems associated with this multilevel DAG level by level, from the lowest level to the highest level, is called a "dynamic programming algorithm". For each subproblem considered by the algorithm, the following information is saved in a table:

1) the value of the optimal solution is saved in the "dynamic programming table";

2) the information needed to reconstruct this solution is saved in the "trace table".

In general, the solution of a DP problem is represented by a recursive equation called the functional equation, optimization equation or cost function. In this equation, the left-hand side is an unknown quantity, and the right-hand side is an expression of the minimum (or maximum) of a set of elements, each expressed as a function of the solutions of the subproblems of the lower levels. The nature of the composition function depends on the problem being solved. The functional equation of the problem of finding the shortest path between a pair of nodes in an acyclic graph is presented in section 3.3.5. If the optimal solution of a problem is determined by composing the optimal solutions of the subproblems and selecting the minimum (or maximum), we say that the formulation of this solution is a DP formulation.

3.3. Discrete DP

Let us study an optimization problem of the form:

$$\sup_{x_t} \left\{ \sum_{t=0}^{T-1} V_t(x_t, x_{t+1}) + V_T(x_T) \right\} \qquad [3.1]$$

subject to the constraints: $x_0 = x$ given, $x_t \in A$, and $x_{t+1} \in \Gamma_t(x_t)$ for every $t = 0, \dots, T-1$. Any such problem is said to have a finite horizon T. The set A is called the state space, Γ_t is a mapping from A into A that models the constraints on the dynamics of the state variable x_t (such that $\Gamma_t(x_t)$ is the set of possible successors of x_t), the functions V_t are the instantaneous payments and finally V_T is the terminal payment. These problems are expressed in the form of a maximization because this is standard practice in economics. But the same theory can also be applied to minimization problems because of the identity $-\sup(.) = \inf(.)$.

We will also consider infinite horizon problems with a discounted criterion of the form:

$$\sup_{x_t} \left\{ \sum_{t=0}^{\infty} \beta^t V(x_t, x_{t+1}) \right\} \qquad [3.2]$$

subject to the constraints: $x_0 = x$ given, $x_t \in A$, and $x_{t+1} \in \Gamma(x_t)$ for every $t \leq 0$. The terms A, Γ and V are interpreted as before, and $\beta \in]0, 1[$ is a discounting factor.

Two natural questions to ask about this problem are whether solutions (i.e. optimal policies) to [3.1] and [3.2] exist and whether there is a useful way to characterize them, ideally with a computational strategy. We will see that by exploiting the recursive structure of this problem we can fairly easily accomplish our goal using the principle of DP, the value function and Bellman's equation, which are largely due to Richard Bellman [BEL 56].

Problems of form [3.2] are frequently encountered in economics in particular, for example in models of growth, inventory management, natural resource exploitation and the environment [LUC 89].

3.3.1. *Bellman's equation and the principle of optimality*

There is a general class of optimization problems that satisfy Bellman's principle of optimality. Any problem whose functional satisfies Bellman's equation belongs to this class.

Consider a dynamic system, here assumed to have discrete time and a finite horizon T, whose state at time t $(0 \leq t \leq T)$ is represented by x_t (scalar value or vector in $\mathbb{R}^d$).

Let $\Gamma(x_t)$ be the set of possible actions when the system is in state x_t. If the action $a_t \in \Gamma(x_t)$ is chosen, the system takes on the state $x_{t+1} = \phi(x_t, a_t)$.

Write $F(x_t, a_t)$ for the cost of the action a_t when the system is in the state x_t. Starting from the initial state x_0, the global cost after T time steps is therefore: $\sum_{t=0}^{T} F(x_t, a_t)$.

Minimizing the global profit means solving Bellman's equation:

$$\begin{aligned} &\min_{a_0,\ldots,a_T} \left\{ \sum_{t=0}^{T} F(x_t, a_t) \mid \forall 0 \leq t \leq T,\ a_t \in \Gamma(x_t) \right. \\ &\left. \text{and } \forall 0 \leq t < T,\ x_{t+1} = \phi(x_t, a_t) \right\}. \end{aligned} \quad [3.3]$$

Write $P_T(x)$ for the minimum cost obtained when arriving at state x after T steps, starting from state x_0. The above equation gives the value of $P_T(x_{T+1})$ (after the final decision a_T, the system arrives at the state x_{T+1}, which is not associated with any cost).

To solve Bellman's equation, we must choose an action a_t to guide the system at each step from the set of possible actions, in such a way as to minimize the global cost. For any final state x, this minimization problem can also be written as:

$$P_T(x) = \min_{a \in \Gamma(y) | x = \phi(y,a)} \left\{ P_{T-1}(y) + F(y, a) \right\}, \quad [3.4]$$

where $P_{T-1}(y)$ denotes the minimum cost that can be achieved in $T-1$ steps leading to state y.

In other words, Bellman's equation satisfies the Bellman principle of optimality. The optimal strategy in T steps leading up to x at time T and passing through y at time $T-1$ contains the optimal strategy leading up to y in $T-1$ steps. We can therefore use DP to solve this problem.

To do this, suppose that $P_0(x)$ is known for every possible value of x. For each possible x, we can now compute:

$$P_1(x) = \min_{a \in \Gamma(y) | x = \phi(y,a)} \left\{ P_0(y) + F(y, a) \right\}, \quad [3.5]$$

then compute $P_2(x), \ldots$ up to $P_T(x)$, which are the values we are interested in. The sequence of choices $a_, a_1, \ldots, a_T$ gives the optimal strategy.

3.3.2. *Approach of the method*

The method proceeds as follows:

– the problem is divided into steps and a decision is made at each step;

– each step is associated with a certain number of states;

– at each step, the decision that is taken transforms the current state into a state associated with the next step;

– once a state is given, the optimal strategy for the remaining steps is independent of the decisions taken at previous steps;

– the search algorithm for the optimal solution begins by finding the optimal strategy for every state at the last step;

– a recurrence relation identifies the optimal strategy at each state of the nth step, in terms of the optimal strategy at each state of the $(n+1)$th step.

In any DP problem, we construct a table analogous to the one shown below at each step:

S \ X_1	$f_n(S, X_1)$ state n+1	$f_n^*(s)$	$x^*(n)$
state n	Cost or distance	Length of optimal path from S to the final state	Best state at step $n+1$ along the final optimal path.

Table 3.1. *Table constructed to solve the problem*

3.3.3. *A few examples of DP*

3.3.3.1. *Example of a mailman*

A mailman decides to travel between different cities that are not familiar to him. His goal is to choose the least dangerous path. To choose his path, he asks an insurance provider for quotes on life insurance policies along each possible route of his journey.

The mailman's journey consists of four steps (see Figure 3.1). He needs to arrive at city J after departing from city A. The edges between these nodes represent the different possible trips that he can take, and the values c_{ij} above the edge represent the price of a life insurance policy along the trip described by the edge. The least dangerous path is the one with the cheapest life insurance policy.

Let x_n $(n = 1, \ldots, 4)$ be the decision variables for each of the four steps. The path followed by the mailman is $A \to x_1 \to x_2 \to x_3 \to x_4$ with $x_4 = J$.

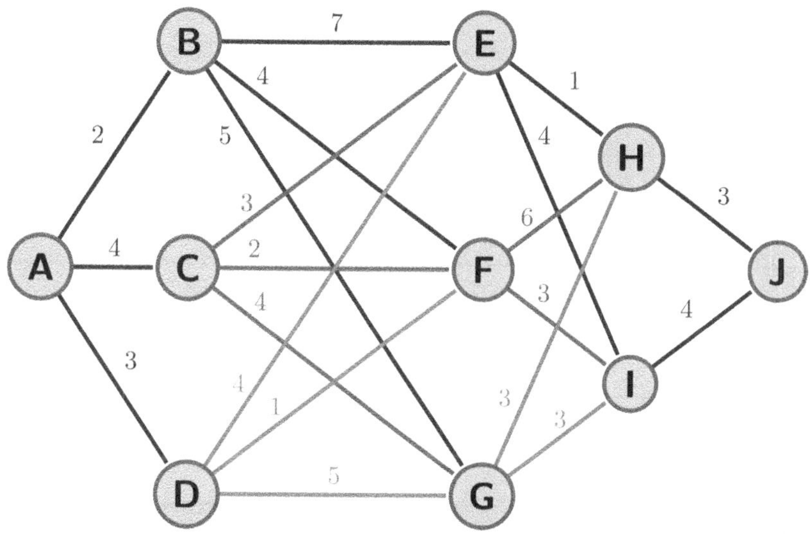

Figure 3.1. *Different cities. For a color version of this figure, see www.iste.co.uk/radi/optimizations.zip*

Let $f_n(S, x_n)$ be the total cost of the life insurance policy for the remaining steps if we are currently at the state S of step n and the chosen destination is x_n.

Step 4:

x_4 \ S	$f_4(S, x_4)$	$f_4^*(S)$	x_4^*
H	3	3	J
I	4	4	J

Step 3:

S \ X_3	$f_3(S, X_3) = C_S X_3 + f_4^*(X_3)$		$f_3^*(S)$	X_3^*
	H	I		
E	$4(= 1 + 3)$	$8(= 4 + 4)$	4	H
F	9	7	7	I
G	6	7	6	H

Step 2:

S \ X_2	$f_2(S, X_2) = C_S X_2 + f_3^*(X_2)$			$f_2^*(S)$	X_2^*
	E	F	G		
B	11	11	12	11	E or F
C	7	9	10	7	E
D	8	8	11	8	E or F

Step 1:

S \ X_1	$f_1(S, X_1) = C_S X_1 + f_2^*(X_1)$			$f_1^*(S)$	X_1^*
	B	C	D		
A	13	11	11	11	C or D

The value of (A) is equal to 11, representing the minimum total cost of the life insurance policy.

The set of these paths is:

$$A \longrightarrow C \longrightarrow E \longrightarrow H \longrightarrow J$$

$$A \longrightarrow D \longrightarrow E \longrightarrow H \longrightarrow J$$

$$A \longrightarrow D \longrightarrow F \longrightarrow I \longrightarrow J$$

A deterministic dynamic problem is characterized by its objective function. This function can, for example, be the minimum of the sum of the contributions that arise when passing from one state to another, or the maximum of such a sum, or the minimum of the product of such terms.

Step n Step $n+1$

$$S_n \quad \overset{\text{contribution of } x_n}{\longrightarrow} \quad S_{n+1}$$

3.3.4. *Solving an LP*

Linear programming can be used to solve linear programs of small sizes. Here, we restrict attention to integer linear programs. Consider the following linear program:

$$\begin{aligned} \min\ & 3x_1 + 5x_2 \\ \text{s.t.}\ & 3x_1 + 2x_2 \leq 9 \\ & 4x_1 + 5x_2 \leq 11 \\ & x_1, x_2 \geq 0 \\ & x_1, x_2 \text{ integers} \end{aligned} \qquad [3.6]$$

The corresponding decision is to determine the values of x_1 and x_2. This problem can be assimilated to a two-step dynamic problem, where we must first determine the value of x_1 followed by the value of x_2. The states corresponding to each step remain to be determined.

An idea that occurs after some straightforward reflection is to define the state as the number of unused resources in the constraints. Each state is therefore described by a vector (r_1, r_2), where r_1 and r_2 denote the unused resources in the first and second constraints, respectively.

At step 1, the only possible state is $S = (9, 11)$. The basic structure of the problem is:

Step n	Step $n+1$

$$(r_1, r_2) \xrightarrow{c_n x_n} (r_1 - a_{1n}x_n, r_2 - a_{2n}x_n)$$

$$f_n((r_1, r_2), x_n) = c_n x_n + f^*_{n+1}(r_1 - a_{1n}x_n, r_2 - a_{2n}x_n),$$

and we have $f^*_3(r_1, r_2) = 0$.

Step 2

At step 2, if we consider every possible combination of the values of r_1 and r_2, we would need to consider 99 states in total. But not all of these states are feasible. For example, the state $(10, 8)$ will never arise in step 2. Therefore, to determine the set of feasible states in the second step, we only need to consider values of x_1 that are possible (0, 1 and 2), which gives us the states $(9, 11)$, $(6, 7)$ and $(3, 3)$.

S \ x_2	$f_2(r_1, r_2) = 5x_2$			$f_2^*(S)$	x_2^*
	0	1	2		
9,11	0	5	10	10	2
6,7	0	5	-	5	1
3,3	0	-	-	0	0

Step 1

We have:

$$f_1((r_1, r_2), x_1) = 3x_1 + f_2(r_1 - 3x_1, r_2 - 4x_1).$$

S \ x_1	$f_1(r_1, r_2)$			$f_1^*(S)$	x_1^*
	0	1	2		
9,11	10	3+5=8	6	10	0

The optimal solution is $(x_1, x_2) = (0, 2)$, and the value of the objective function is 10.

3.3.5. *Shortest path problem*

Consider the problem of finding the shortest path in a directed and weighted graph, defined on a set of n vertices $\{1, 2, \ldots, n\}$. If the weights are all positive, Dijkstra's algorithm allows us to find the shortest path (i.e. the path with minimum total weight) between two vertices.

If the weights are allowed to be negative but the graph does not have a cycle of negative total weight, the Floyd–Warshall algorithm presented below gives an answer to this question.

Let $d(i, j)$ be the weight of the edge (i, j), and (for $k \geq 1$) let $d_{i,j}^k$ be the length of the shortest path from vertex i to vertex j among paths that only pass through the vertices $\{1, 2, \ldots, k\}$. By convention, we will also assume that:

$$\begin{cases} d_{i,j}^k = +\infty, & \text{if there is no path from } i \text{ to } j \\ d_{i,j}^0 = d(i,j) \end{cases} \qquad [3.7]$$

The goal is to construct an algorithm that allows us to compute the $d_{i,j}^n$.

Suppose that k satisfies $0 \leq k \leq n-1$. The $(d_{i,j}^{k+1})_{i,j}$ satisfy the following recursion relation with the $(d_{i,j}^{k})_{i,j}$:

$$\forall (i,j) \in \{1,2,\ldots,n\}^2,\ d_{i,j}^{k+1} = \min\{d_{i,j}^{k}, d_{i,k+1}^{k} + d_{k+1,j}^{k}\}. \qquad [3.8]$$

The Floyd–Warshall(–Roy) algorithm proceeds by computing the $(d_{i,j}^{1})_{i,j}$ (in $O(n^2)$ operations, since there are n^2 pairs of vertices (i,j)), then the $(d_{i,j}^{2})_{i,j}$ ($O(n^2)$ operations), and finally the $(d_{i,j}^{n})_{i,j}$ (once again $O(n^2)$ operations) that yield the desired paths of shortest length. The overall complexity of the algorithm is $O(n^3)$.

Consider the problem of finding the shortest paths from a vertex s_0 in a directed acyclic graph G whose weights $d(x,y)$ are allowed to be negative. Let $d_T(s)$ be the length of the shortest path in T steps arriving at s. Then:

$$d_T(s_T) = \min_{s_1,\ldots,s_T} \left\{ \sum_{t=1}^{T} d(s_{t-1}, s_t)\ \forall 1 \leq t \leq T,\ s \in G(s_{t-1}) \right\}, \qquad [3.9]$$

where $G(s)$ denotes the set of successors of a vertex s of the graph G.

This can be recognized as Bellman's equation. The corresponding dynamic system involves traveling across a graph, beginning at vertex s_0 while minimizing the total weight $d_T(s)$. Performing an action in a state y yields a successor x of y with a cost $d(x,y)$.

The problem can therefore be solved with DP by recursively finding:

$$d_T(x) = \min_{y \in G^{-1}(x)} \{d(x,y) + d_{T-1}(y)\}. \qquad [3.10]$$

We want to find $d_T(s_0)$. Since the graph is acyclic, we can restrict attention to $T \leq n$.

The algorithm proceeds by applying this recursion formula with the initialization: $d_0(s_0) = 0$ and $d_0(u) = +\infty$ if $u \neq s_0$. The length of the shortest path starting from s_0 and arriving at a vertex u is given by $\min_{0 \leq t \leq n} d_t(u)$. This idea is the basis of the Bellman–Ford algorithm, which finds the shortest path between a given vertex and any other vertex of a weighted graph.

The DAG, modeling the dependency relation between the subproblems (subpaths), is shown in Figure 3.2. This DAG can be organized into a multilevel DAG in several ways, some of which are level dependent (Figure 3.3), and at least one of which is level independent. (Figure 3.4).

EXAMPLE 3.1.– Suppose that we wish to determine the shortest path from A to F in the following graph (see Figure 3.5).

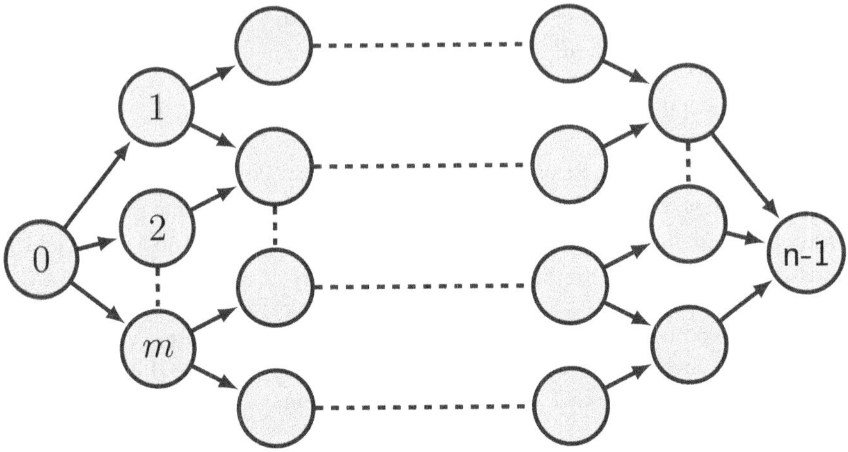

Figure 3.2. *DAG representing the dependencies between the different subproblems of the shortest path problem. For a color version of this figure, see www.iste.co.uk/radi/optimizations.zip*

For $k = 1$, we have:

$$F_1(B) = 2P_1(B) = A$$
$$F_1(D) = 6P_1(D) = A$$
$$F_1(E) = 5P_1(E) = A$$

For the other values of k, we have: $F_2(C) = F_1(B) + 1 = 3$;

$$F_2(F) = \min(F_1(F), F_1(B) + 7, F_1(D) + 6, F_1(E) + 1)$$
$$= \min(\infty, 9, 12, 6) = 6$$
$$F_3(E) = \min(F_2(E), F_2(A) + 5, F_2(C) + 1)$$
$$= \min(5, 5, 4) = 4$$
$$F_4(F) = \min(F_3(F), F_3(B) + 7, F_3(D) + 6, F_3(E) + 1)$$
$$= \min(6, 9, 12, 5) = 5$$

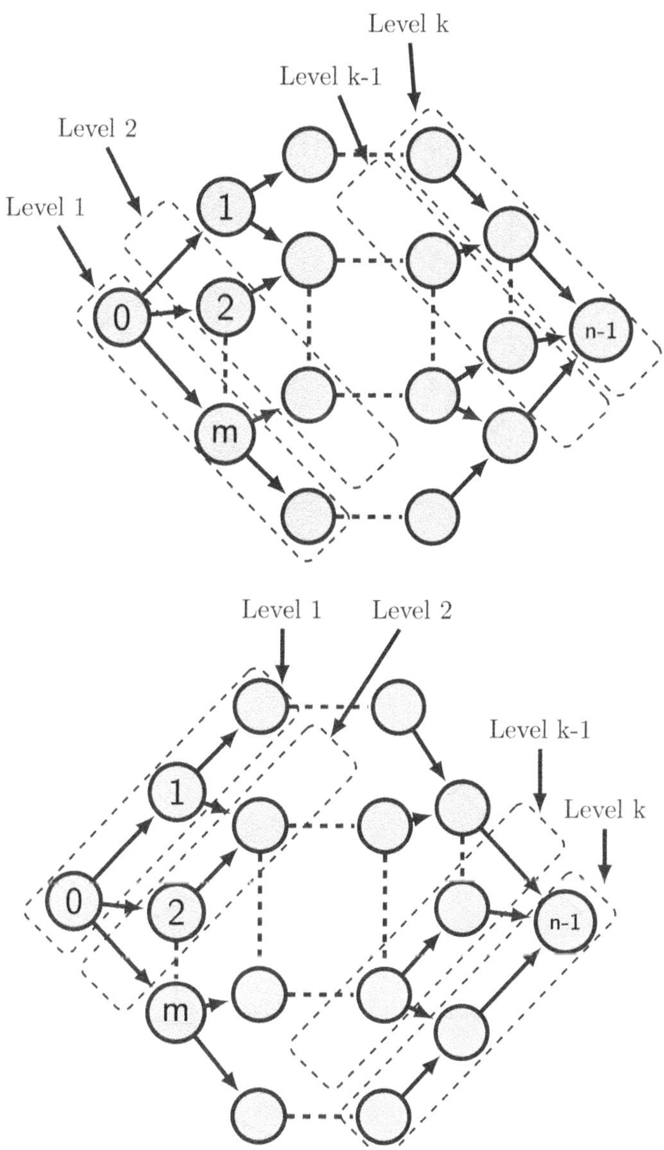

Figure 3.3. *Level-dependent multilevel DAG for the shortest path problem. For a color version of this figure, see www.iste.co.uk/radi/optimizations.zip*

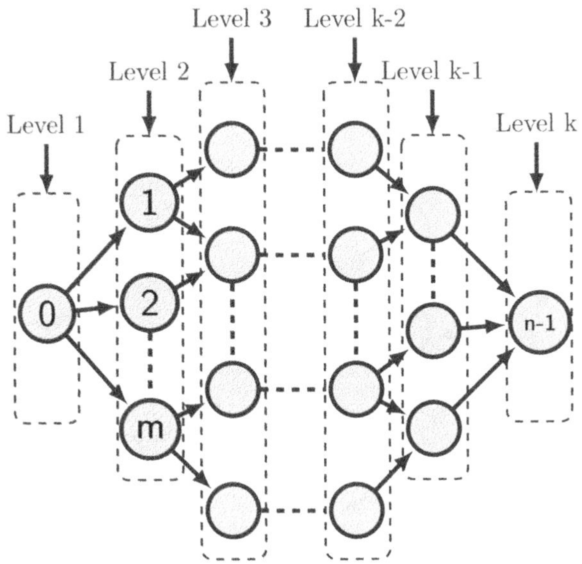

Figure 3.4. *Level-independent multilevel DAG for the shortest path problem. For a color version of this figure, see www.iste.co.uk/radi/optimizations.zip*

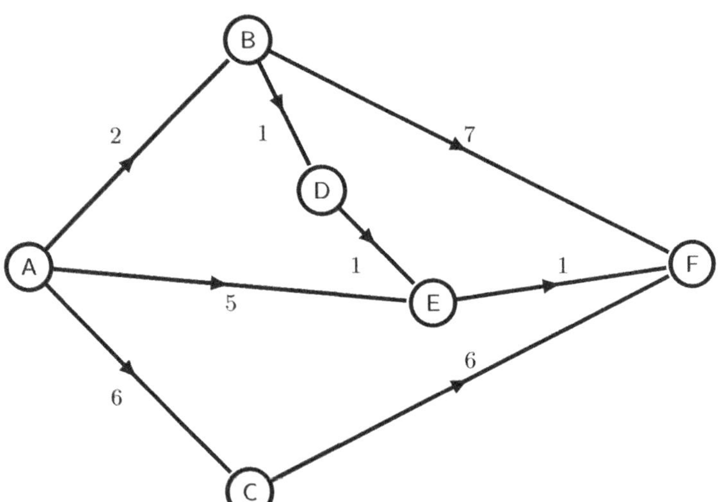

Figure 3.5. *Graph corresponding to this problem*

We therefore have:

$$X_4 = F$$

$$X_3 = P_4(F) = F$$

$$X_2 = P_3(E) = C$$

$$X_1 = P_2(C) = B$$

$$X_0 = P_1(B) = A$$

The optimal path (A, B, C, E, F) of minimum length is 5.

3.3.6. *Knapsack problem*

Suppose that we have n objects with utilities $c_1, \ldots, c_n$ and weights (or volumes) $w_1, \ldots, w_n$. The goal is to fill the knapsack while maximizing $\sum_{i \in I} c_i$ subject to the constraint $\sum_{i \in I} w_i \leq W$, where W is the maximum capacity of the knapsack.

To solve the problem recursively, we first make it more complicated. A more complex filling problem than the initial problem is one where the size of the knapsack and the number of objects being considered are arbitrary. We then define $C(v, i)$ as the expression of the best cost for filling a knapsack of size v with the first i objects. Solving this knapsack filling problem of size W with n objects means finding the value of $C(W, n)$.

The recursion: the last object is either chosen or not chosen, so:

$$C(v, i) = \max \begin{cases} C(v, i-1), & \text{final object is not considered} \\ C(v - w_i, i-1) + c_i, & \text{final object is chosen} \end{cases} \qquad [3.11]$$

The optimal solutions of the subproblems are used to find the optimal solution of the global problem without ever computing the same value twice. This requires the dependencies of the computation to be taken into account.

EXAMPLE 3.2.– Consider the knapsack filling problem (without repetition): given a list of n objects with weights $p_1, \ldots, p_n$ and utilities $v_1, \ldots, v_n$, how can we fill a knapsack with a maximum weight capacity of P in such a way that the utility is maximized?

Let us find the composition with the highest utility for a knapsack of capacity $P = 5$ for the list of objects whose weights and utilities are given in Table 3.2.

j	1	2	3	4
p_j	2	1	2	3
v_j	3	2	1	4

Table 3.2. *List of n objects*

We define the following subproblem: let j satisfy $0 \leq j \leq n$, and let $V(p, j)$ be the maximum utility that can be placed in a knapsack with maximum capacity p by selecting objects from among $1, \ldots, j$. We want to find the value of $V(P, n)$. The relation between $V(., j)$ and $V(., j-1)$ is as follows: if $j \geq 1$ and $p \geq p_j$, then:

$$V(p, j) = \max\{V(p - p_j, j-1) + v_j, V(p, j-1)\},$$

and, if $p < p_j$, then $V(p, j) = V(p, j-1)$.

Indeed, a knapsack with capacity $p < p_j$ cannot contain the jth object, and the same maximum utility is found by restricting the choice to the first $j-1$ objects. And if $p \geq p_j$, either the maximum utility $V(p, j)$ is achieved by choosing the jth object, then the rest of the knapsack is filled in the same way as a knapsack of maximum capacity $p - p_j$ using the $j-1$ first objects (in which case $V(p, j) = V(p - p_j, j-1) + v_j$), or this value is achieved without taking the jth object, and therefore by making a selection from among the first $j-1$ objects (in which case $V(p, j) = V(p, j-1)$). The maximum utility of the knapsack can therefore be computed recursively with the initialization $V(p, 0) = 0$ and $V(0, j) = 0$ by filling out a table.

To apply this numerically, we must fill out Table 3.3 using the recursive formula established earlier (for increasing j then with fixed j for increasing p).

p \ j	0	1	2	3	4
0	0	0	0	0	0
1	0	0	2	2	2
2	0	3	3	3	3
3	0	3	5	5	5
4	0	5	5	5	6
5	0	3	5	6	7

Table 3.3. *Results of the recursive formula*

Backtracking then allows us to conclude that the optimal knapsack (which has utility 7) consists of the objects 4 and 1.

Another way to solve this problem is to formulate a Bellman's equation for it. Here, the goal is to maximize $\sum_{j=1}^{n} \delta_j v_j$, where δ_j is 1 if the object j is placed in the knapsack and 0 otherwise, subject to the constraint: $\sum_j \delta_j p_j \leq P$. Note that this is an integer linear programming problem.

The graph is then constructed from the decisions taken after starting with an empty knapsack (capacity 5). The remaining capacity after making a decision at each level $< j$ is assigned to the vertices of the graph of level j, and the weights of the edges are v_j. This is shown in Figure 3.6.

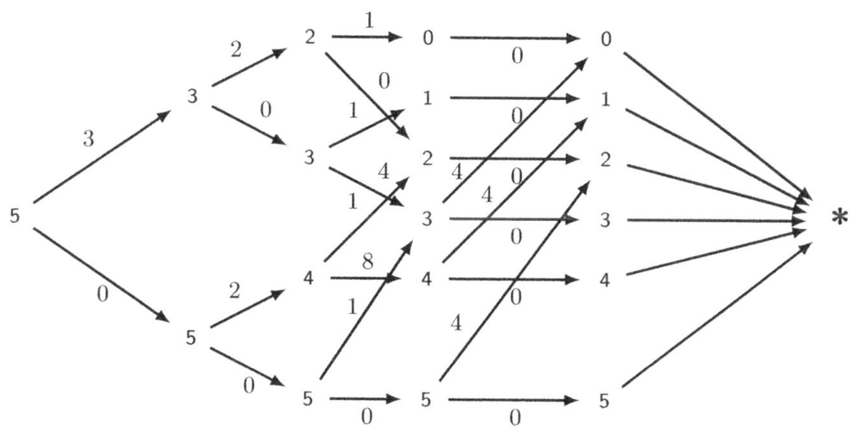

Figure 3.6. *Graph constructed from this problem*

Solving this problem is thus equivalent to finding the longest path between the leftmost vertex and the rightmost vertex. The solution is of course the same as the solution found by the table method.

3.3.7. *Stock management problem*

Stock management problems are classical examples in DP. This is partly due to the fact that they can be immediately, and very naturally, expressed with its formalism.

Consider a company that manufactures and sells a product M. To manufacture this product, the company requires a specific part P that is purchased from a supplier. The company has a warehouse that can store up to 5 parts P. At the start of each month, it buys some quantity of parts P to meet the demands of the month. The purchase price of the parts P fluctuates from month to month but is assumed to be known in advance. Furthermore, the company knows the demand of product M for each period being considered. This information (purchase price P and demand of M to be satisfied) is presented in Table 3.4.

	Month 1	Month 2	Month 3	Month 4	Month 5
Unit purchase price p_i	3	2	2	4	3
Demand d_i	1	4	2	3	1

Table 3.4. *Demand and purchase prices*

Given that one part P is needed to manufacture one object M, the goal of the problem is to determine an optimal purchasing strategy for the company; more precisely, we wish to decide the quantity P to purchase at the start of each month in such a way that the demand is satisfied and the total cost is minimized. We also assume that the initial stock is zero and that the final stock should also be zero.

The way to partition the problem into periods is obvious. The ith month corresponds to the period i, and we add an initial period 0. The state of period i is the stock at the end of the period. Thus, there is one state $s_0 \in S_0$, corresponding to a stock of zero, six states for each of the periods one to four, corresponding to a stock of between 0 and 5, and one state at period five, corresponding to a final stock of zero.

The decisions are the quantities of parts P to purchase at the start of each period. If we have a stock k_{i-1} at the end of period $i-1$, we can purchase at most $l_i = 5 - k_{i-1}$ parts without exceeding the storage capacity. Additionally, we need at least d_i parts to satisfy demand. This determines the decisions that are possible in each state, i.e. the set A: given a stock k_{i-1} at period $i-1$,

$$(k_{i-1}, l_i) \in A \Leftrightarrow k_{i-1} + l_i \in \{d_i, d_{i+1}, \dots, 5\}.$$

The transition function is the computation of the stock k_i at the end of period i as a function of the stock k_{i-1} at the end of period $i-1$ and the quantity l_i purchased at the start of period i. Thus, $k_i = k_{i-1} + l_i - d_i$. There is also a very natural way to value each policy. The goal is to minimize the sum of the costs of purchasing raw material. Therefore, consider a realizable policy that purchases l_i parts P at period i, l_{i+1} parts at period $i+1, \dots, l_j$ at period j. The value of this policy is defined as the cost incurred by these purchases:

$$v(l_i, \dots, l_j) = \sum_{m=i}^{j} p_m l_m.$$

This valuation (which takes real values) is clearly a monotone function.

In this problem, the goal is to determine how many parts P need to be purchased at the start of each month, in order to minimize the total cost (while satisfying the demand). DP proceeds by solving the following family of problems: for each period i and each stock at the end of this period, determine an optimal policy that arrives at this stock. In Table 3.5, the column index corresponds to the period being considered, and the row index corresponds to the stock at the end of the period. The cells show the value of an optimal policy that reaches this state and the decision vector that defines it.

Stock \ Period	0	1	2	3	4	5
0	**0** ()	**3** **(1)**	**11** **(1,4)**	15 (1,4,2)	**21** **(1,4,5,0)**	**24** **(1,4,5,0,1)**
1		6 (2)	13 (1,5)	17 (1,4,3)	25 (1,4,5,1)	
2		9 (3)		19 (1,4,4)	29 (1,4,5,2)	
3		12 (4)		**21** **(1,4,5)**		
4		15 (5)				
5						

Table 3.5. *Solution of the stock problem*

3.4. Continuous DP

Consider the value function defined as follows:

$$v(t,x) = \sup\left\{ \int_t^T L(s, y(s), \dot{y}(s))ds + g(y(T)) \mid y \in \mathcal{C}^1([t,T], \mathbb{R}^n)\ y(t) = x \right\}. \qquad [3.12]$$

The function v satisfies the boundary condition $(T, x) = g(x)$ for every $x \in \mathbb{R}^n$.

The principle of DP states that if a curve $y(.)$ starting from x at $t = 0$ is optimal between 0 and T, then it is also optimal between t and T among all curves with the value $y(t)$ at time t. This principle is expressed by the following relation.

For every $x \in \mathbb{R}^n$ and all $t \in [0, T]$, the value function satisfies:

$$v(0,x) = \sup\left\{ \int_0^t L(s, y(s), \dot{y}(s))ds + v(t, y(t)) \mid y(0) = x \right\}.$$

3.4.1. *Hamilton–Jacobi equation*

Another property of v is that it is the solution of a first-order partial differential equation (PDE) called the Hamilton–Jacobi equation:

THEOREM 3.1.– Suppose that v is regular. Then v is a solution of the Hamilton–Jacobi equation:

$$\partial_t v(t,x) + H(t,x,\nabla_x v(t,x)) = 0, \qquad [3.13]$$

where H is the Hamiltonian, defined for every $p \in \mathbb{R}^n$ by:

$$H(t,x,p) = \sup_{q \in \mathbb{R}^n} \{q.p + L(t,x,q)\}. \qquad [3.14]$$

In addition, it would be desirable to characterize v as the unique solution of the Cauchy problem:

$$\partial_t v(t,x) + H(t,x,\nabla_x v(t,x)) = 0$$
$$v(T,.) = g(.)$$

To establish results like this without making the unrealistic hypothesis that v is regular, we need to appeal to the notion of viscosity solution [BAR 94].

3.4.2. *Application to a consumption-savings model*

This application is taken from the lecture notes in [CAR 07]. Consider the context of continuous time over period $[0,T]$, and consider a household whose wealth, instantaneous salary, consumption and savings are denoted as $x(t)$, $S(t)$, $c(t)$ and $e(t)$, respectively. Finally, suppose that the household wishes to maximize the following utility function:

$$\int_0^T (-\delta t)\log(c(t))dt + e^{-\delta T} V(x(T)).$$

We have the relations:

$$S(t) = c(t) + e(t), \ \dot{x}(t) = e(t) + rx(t),$$

where r is the exogenous interest rate (assumed to be constant as a simplification). Since the initial wealth x_0 of the household is given, the choice of optimal consumption savings for this household reduces to the variational problem

$$\sup J(x) = \int_0^T e^{-\delta t} \log(S(t)+rx(t)-\dot{x}(t))dt + e^{\delta T} V(x(T))dt \colon x(0) = x_0.$$

If we further assume that V is concave, then first-order conditions are sufficient to guarantee optimality. After setting $c(t) = S(t) + rx(t) - \dot{x}(t)$, the Euler–Lagrange equation states that:

$$-\frac{d}{dt}\left(\frac{e^{-\delta t}}{c(t)}\right) = \frac{re^{-\delta t}}{c(t)}.$$

Setting $y(t) = e^{-\delta t}/c(t)$ therefore gives $y(t) = e^{-rt}/c(0)$, and hence:

$$c(t) = e^{(r-\delta)t}c(0).$$

We still need to determine the constant $c(0)$. To do this, we need to return to the (state) variable x by integrating $\dot{x} - rx = S - c$, and then apply the transversality condition at T. In this case, it can be stated as follows:

$$V'(x(T)) = \frac{1}{c(T)}.$$

3.5. Stochastic DP

In most real problems, we need to account for uncertainty. In such cases, the process no longer only involves choices by a decision-maker, but also an element of uncertainty linked to external parameters. This requires us to adapt the deterministic framework considered up to this point. For example, in a stock management problem, the demand is rarely deterministic. Nevertheless, statistical data can be used to model the uncertainty as a random variable corresponding to the demand. The sequence of choices is then studied from the perspective of a given decision criterion.

3.5.1. *Decision-chance process*

In this new context, each period is divided into two consecutive phases: a decision, followed by a random event that determines the state of the system at the end of the period. Given a state and a decision, this random event is defined by specifying a set of states that can be occupied and the probability that each of them will effectively be occupied. This is equivalent to considering a transition function that is random rather than deterministic.

A policy no longer consists of a sequence of decisions, which no longer makes sense in the presence of chance. Instead, for each state $s \in S$, we must specify the decision $x(s)$. The decision-chance process is often represented by a particular type of graph to summarize the states, called the decision-chance graph, which distinguishes between two types of vertex:

– vertices $s_i \in S_i$ corresponding to different states;

– vertices $h_i \in H_i$ corresponding to random events.

In this graph, each edge (s_i, h_i) corresponds to a decision, and the set of edges

$$\{(h_i, s_{i+1}) \mid s_{i+1} \in \Gamma^+(h_i)\}$$

corresponds to the set of possible random transitions. A transition probability $P(h_i, s_{i+1})$ is assigned to each edge (h_i, s_{i+1}).

In such a context, the final possible state is often not unique. Instead, there is a set of final states S_n that might be reached. A scalar value $v(s_n)$ is assigned to each final state $s_n \in S_n$. The value of a policy δ is then the expected value:

$$E(\delta) = \sum_{s_n \in S_n} P_\delta(s_n)v(s_n), \qquad [3.15]$$

where $P_\delta(s_n)$ denotes the probability of reaching the state s_n by following the policy δ.

Depending on whether the values assigned to the final state represent costs or profits, the goal is either to minimize or to maximize this expected value.

3.5.2. *Solving method*

The value of an optimal policy is obtained by starting from the final states and working backward to the initial state using the following recurrence relations:

$$v(s_i) = \min\{v(h_i) \mid h_i \in \Gamma^+(s_i)\} \qquad [3.16]$$

$$v(h_i) = \sum_{s_{i+1} \in \Gamma^+(h_i)} P(h_i, s_{i+1})v(s_{i+1}) \qquad [3.17]$$

A label is assigned to each vertex of the decision graph containing the value of an optimal policy from this vertex to a final state. An optimal decision $x^*(s_i)$ at a state s_i is an edge (s_i, h_i) for which the label assigned to the vertex h_i is minimal. The set of these optimal decisions is an optimal policy.

3.5.3. *Application to a contract problem*

Consider an insurance contract that allows the buyer to subscribe for year 1 and/or year 2. Of course, the buyer does not know whether he will need to activate the contract after an accident. Furthermore, the probability of having an accident in year 2 depends conditionally on whether or not there was an accident in year 1.

The decision-chance tree for this problem is shown in Figure 3.7. The vertex d_1 represents the choice of whether to subscribe to the insurance contract (branch 1) or not (branch 0) in the first year. The vertices h_1 and h_2 represent whether an event requiring the contract to be activated occurs (branch 1) or not (branch 2) during the first year ($P(0) = \frac{2}{5}$ and $P(1) = \frac{3}{5}$). The vertices d_2, d_3, d_4 and d_5 represent the choice of whether or not to subscribe to the insurance contract in the second year. Finally, the vertices h_3, h_4, h_5, h_6, h_7, h_8, h_9 and h_{10} represent whether or not an event requiring the contract to be activated occurs during the second year. If an accident already occurred in the first year, the probability of such an event in the second year is $P(0) = \frac{3}{4}$ (and so $P(1) = \frac{1}{4}$). However, if there was no accident in the first year, the probabilities remain as $P(0) = \frac{2}{5}$ and $P(1) = \frac{3}{5}$.

The total cost incurred is shown on each leaf of the decision-chance tree. Subscribing to a contract costs 1, and experiencing an accident without coverage costs 2 (as opposed to 0 with coverage). It is straightforward to verify that these parameters lead to a value of $\min\{\frac{6}{5}, 1\} = 1$ at vertex d_2, a value of $\frac{5}{2}$ at vertex d_3, a value of 2 at vertex d_4 and a value of $\frac{3}{2}$ at vertex d_5.

Therefore, the label of vertex h_1 has the value

$$1 \times \frac{2}{5} + \frac{5}{2} \times \frac{3}{5} = \frac{19}{10},$$

and the label of vertex h_2 has the value $\frac{17}{10}$. The optimal policy is therefore to subscribe to the contract the first year and only subscribe to the contract the second year if there was no accident in the first year.

3.5.4. *Optimal binary search tree*

Finding the optimal binary search tree (OBST) is a widely studied problem in the field of operations research.

DEFINITION 3.1.– *A binary search tree is a binary tree that allows the ordered set of elements that make up the tree to be efficiently searched.*

Any such tree gives a complete ordering of its elements as follows:

– every value less than the value of a node is found in the left subtree of this node;

– every value greater than the value of a node is found in its right subtree.

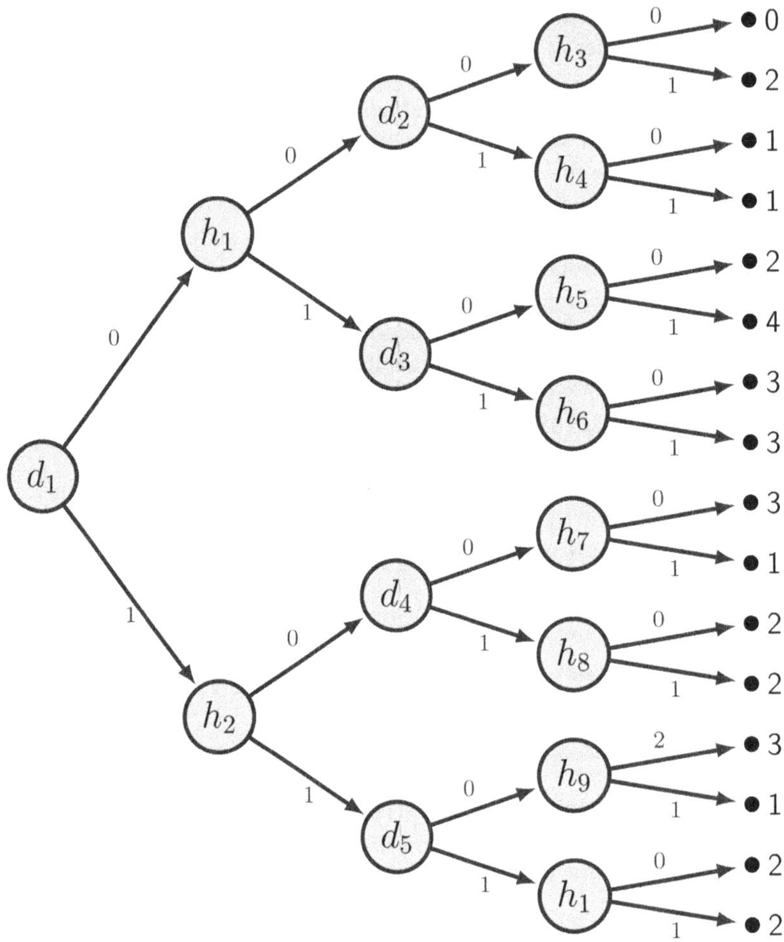

Figure 3.7. *Decision-chance tree for the insurance contract problem. For a color version of this figure, see www.iste.co.uk/radi/optimizations.zip*

In a binary search tree, the cost of finding an element depends on its depth in the tree. Consider the set of words $A = \{$all, bas, cet, des, elle, for, hola, idem, joli, le, ma, non, pour, que, tel$\}$, ordered alphabetically, as shown in Figure 3.8. The cost of finding the element *le* is 1, whereas the cost of finding the element *hola* is 5.

It is easy to see that:

– when sorting a tree of n elements, it is possible to construct $\Omega\left(\frac{4^n}{n^{3/2}}\right)$ different binary trees [KNU 71]. Figure 3.9 presents the different possible trees containing the list of words S={le, la, les};

– the cost of finding an element k depends on the tree used for the search.

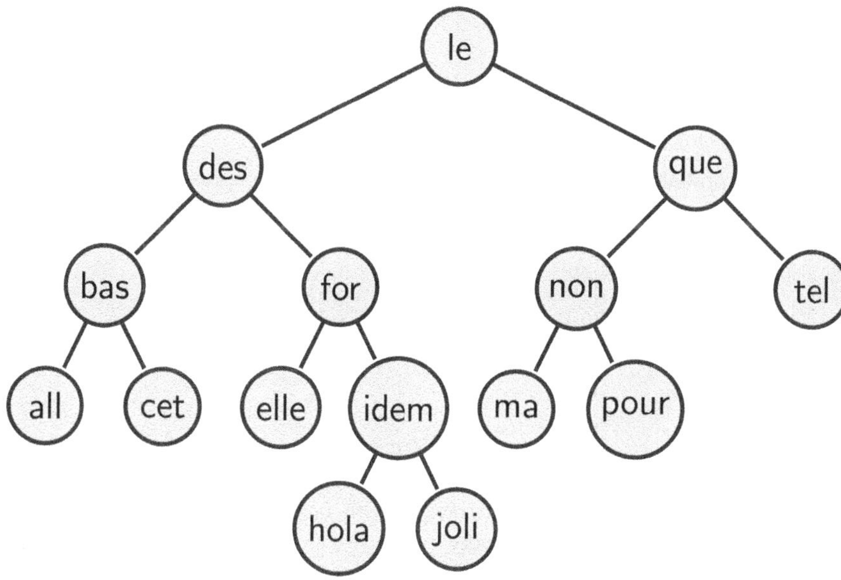

Figure 3.8. *Binary search tree for set A. For a color version of this figure, see www.iste.co.uk/radi/optimizations.zip*

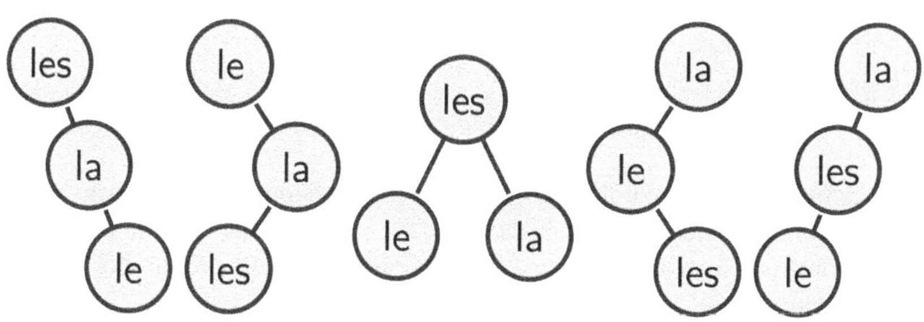

Figure 3.9. *Five possible binary trees for list S. For a color version of this figure, see www.iste.co.uk/radi/optimizations.zip*

3.5.4.1. *Formulation of the problem*

Consider a set of n elements $k_1 < k_2 < \ldots < k_n$ to be arranged into a binary search tree together with a set of $2n+1$ probabilities $p_1, p_2, \ldots, p_n, q_0, q_1, q_2, \ldots, q_n$ satisfying

$$\sum_{i=1}^{n} p_i + \sum_{i=0}^{n} q_i = 1,$$

where:

– p_i $(1 \leq i \leq n)$ is the probability that the target value is equal to k_i in the search process;

– q_i is the probability that the target value in the search process is:

- in the interval $]k_i, k_{i+1}[$ if $0 < i < n$;
- less than k_1 if $i = 0$;
- greater than k_n if $i = n$.

Now, let:

– T be the binary search tree consisting of this set of elements. The set $\{f_0, f_1, \ldots, f_n\}$ of fictitious leaves is added to the tree T, where f_j $(0 < j < n)$ represents the values of the interval $]k_i, k_{i+1}[$, with f_0 and f_n representing values less than k_1 and greater than k_n, respectively.

– b_i be the number of edges in the path from the root node to the internal node k_i;

– a_i be the number of edges in the path from the root node to the leaf f_i.

The average search cost of a value in tree T is given by the expression:

$$\text{cost}(T) = \sum_{0<i\leq n} p_i \times (b_i + 1) + \sum_{0\leq j\leq n} q_j \times a_j,$$

where $b_i + 1$ is the cost required to reach the element k_i, and the cost of f_j is a_j.

The optimal binary search tree for this set of elements is a binary tree with minimum average search cost. Thus, the problem of finding the optimal binary search tree is to construct such a tree given a set of elements and the probability of needing to access each of them.

3.5.4.2. *Formulation of the solution*

We write:

– obst(i, j) for the optimal binary search tree whose elements correspond to the subset $\{k_{i+1}, \ldots, k_j\}$;

– Tree(i, j) for the cost of constructing this tree.

Let

$$\begin{cases} w(i,j) &= p_{i+1} + \cdots + p_j + q_i + q_{i+1} + \cdots + q_j \\ w(i,i) &= \qquad\qquad q_i. \end{cases}$$

It is easy to observe that $w(0, n) = 1$ and that the cost Tree(i, j) satisfies the following recurrence relation:

$$\text{Tree}(i,j) = \begin{cases} g_i, & \text{if } 0 \leq i = j \leq n \\ \min_{i<k\leq j} \text{Tree}(i, k-1) + \text{Tree}(k, j) + w(i,j), & \text{if } 0 \leq i < j \leq n. \end{cases}$$

The cost of finding the optimal binary search tree is Tree$(0, n)$.

3.6. Using Matlab

Matlab offers DP functionality within the framework of graph theory. It has built-in functions to find the shortest path in a graph with the following syntax:

```
Graph::shortestPathSingleSource(G, StartVertex,
<EndVertex = v>, <SearchWith = Dijkstra | Bellman>,
<SearchFor = Weights | Costs>, <ReturnAsGraph>).
```

David Gleich also offers the downloadable MatlabBGL library, which complements Matlab's suite of algorithms. This library contains a rich collection of algorithms for working with graphs, such as those encountered in graph theory. The MatlabBGL package uses Matlab's native sparse matrix as a graph with algorithms that work in practice.

4

Stochastic Programming

4.1. Introduction

This chapter introduces mathematical concepts for modeling optimization problems involving randomness Such problems are called stochastic optimization problems. Randomness can be present in both the criterion being optimized and the constraints of the problem. We begin by presenting the two major classes of this type of problem: open loop problems, where decisions can be taken in advance before the system evolves, and closed loop problems, where decisions are taken over time and depend on observations about the system as it evolves [KAL 94].

Stochastic programming can also be viewed as mathematical programming with random parameters. With this approach:

– any random events are approximated with scenarios;

– the decisions are classified into steps;

– the constraints must be satisfied with some probability p (for example, $p = 0.95$).

The scenario-based approach appears to provide a reasonable method for generating uncertainty in many cases, although it is not the only possible approach. It assumes that there are finitely many observations that can occur; in other words, the random events follow a discrete distribution. Each possible realization corresponds to one scenario.

Stochastic optimization methods are based on probabilistic and random transition mechanisms. Accordingly, executing these methods several times may lead to different results, even when the optimization problem has the same initial conditions [HAJ 03]. These methods are good at finding the global optimum of the problem. Unlike most deterministic methods, they do not require a starting point, nor any knowledge of the gradient of the objective function to find the optimal solution. They are of order zero. However, they require the objective function to be evaluated a large number of times.

In the following, all random variables are written in bold.

4.2. Presentation of the problem

Let $\mathbf{U}$ and $\mathbf{V}$ be two random variables defined on a probability space $(\Omega, A, \mathbb{P})$ taking values in the Hilbert spaces $\mathbb{U}$ and $\mathbb{V}$. We will also write $\mathcal{U} = L^2(\Omega, A, \mathbb{P}; \mathbb{U})$ and $\mathcal{V} = L^2(\Omega, A, \mathbb{P}; \mathbb{W})$. Thus, we are considering the spaces of square-integrable random variables. The goal is to define a few optimization algorithms that take advantage of the notion of gradient that can be naturally defined in a Hilbert space. Let $j \colon \mathcal{U} \times \mathcal{V} \to \mathbb{R}$ be a normal integrand. In general, the stochastic optimization problem can be written as follows:

$$\min_{\mathbf{U} \in \mathcal{U}^{ad}} J(\mathbf{U}) := \mathbb{E}(j(\mathbf{U}, \mathbf{V})), \qquad [4.1]$$

where the admissible set $\mathcal{U}^{ad}$ is a subset of $\mathcal{U}$. Constraints on the random variable $\mathbf{U}$ will be defined on this set.

In some cases, the function J is not known analytically and must be derived from j. Consider a σ-algebra $\mathcal{G}$ on Ω and an A-measurable multifunction $\Gamma \colon \Omega \rightrightarrows \mathcal{P}(\mathbb{U})$. In the following, we will often distinguish between two types of constraint:

– so-called point or almost sure constraints:

$$\mathcal{U}^{ps} := \{\mathbf{U} \in \mathcal{U},\ \mathbf{U} \in \Gamma,\ \mathbb{P} - a.s.\}, \qquad [4.2]$$

– and measurability constraints:

$$\mathcal{U}^{mes} := \{\mathbf{U} \in \mathcal{U},\ \mathbf{U} \text{ is } \mathcal{G} - \text{measurable}\}. \qquad [4.3]$$

$\mathcal{U}^{mes}$ is also a vector subspace of $\mathcal{U}$. Thus, we set: $\mathcal{U}^{ad} = \mathcal{U}^{ps} \cap \mathcal{U}^{mes}$, and, instead of "$\mathbf{U}$ is $\mathcal{G}$-measurable", we often write: $\mathbf{U} \succeq \mathcal{G}$.

The concept of measurability constraint is particular to stochastic optimization. Since each element of the σ-algebra $\mathcal{G}$ can be interpreted as a piece of information, measurability constraints represent the fact that the decision can depend on observations of the system.

4.3. Optimal feedback in an open loop

Suppose that the σ-algebra $\mathcal{G}$ does not depend on $\mathbf{U}$. Even in this case, there are various possibilities for $\mathcal{G}$. The most straightforward is as follows:

DEFINITION 4.1.– *We say that the problem [4.1] is an open loop problem if:* $\mathcal{U}^{mes} \subset \{\mathbf{U} \in \mathcal{U} \mid \mathbf{U}$ *is* $\sigma\{\emptyset, \Omega\}$*-measurable}.*

If not, the problem is said to be a closed-loop problem. Thus, open-loop problems are the case where $\mathcal{G}$ is the trivial σ-algebra $\sigma\{\emptyset, \Omega\}$. They model situations where the decision is taken without any information. In other words, all decisions must be taken from the very start, before any randomness is introduced. The goal is to find the best compromise across every possible random possibility.

Conversely, closed-loop problems describe cases where the decision-maker is able to observe all or some random occurrences and incorporate these observations into his decisions.

In general, open-loop problems are less complex than closed-loop problems, due to the nature of the optimization variables:

– with open loops, we are looking to define *a priori* decisions;

– with closed loops, we are looking to define a decision for every possible observation of the system.

We can often use tools similar to those from deterministic optimization, such as the stochastic gradient method [ROB 51, BER 00] (which applies a stochastic gradient to an optimal investment problem) to tackle open-loop problems. If the problem is large, we can also use the same decomposition techniques as in the deterministic case [COH 90].

4.4. Stochastic linear programming

A linear program whose coefficients are realizations of random variables is called a stochastic linear program. This program targets a certain objective while accounting for a probability function associated with various scenarios. It has been shown that implementations of linear stochastic models are necessarily limited, due to problems involving combinatorial explosion and the fact that only a limited number of scenarios can be taken into account.

EXAMPLE 4.1.– Consider the following LP:

$$\begin{aligned} \min\ & -x_2 \\ \text{s.t.}\ & x_1 + x_2 + x_3 = 2 \\ & \alpha_1.x_1 + \alpha_2.x_2 + x_4 = 2 \\ & -1 \leq x_1 \leq 1,\ x_j \geq 0,\ j = 2, 3, 4 \end{aligned} \qquad [4.4]$$

Set $y_1 = x_1 + 1$, i.e. $x_1 = y_1 - 1$. The LP 4.4 then becomes:

$$\begin{aligned} \min\ & -x_2 \\ \text{s.t. }\ & y_1 + x_2 + x_3 = 3 \\ & \alpha_1.y_1 + \alpha_2.x_2 + x_4 = 2 + \alpha_1 \\ & 0 \leq y_1 \leq 2,\ x_j \geq 0,\ j = 2,3,4 \end{aligned} \quad [4.5]$$

Returning to classical notation in terms of x_i, this gives

$$\begin{aligned} \min\ & -x_2 \\ \text{s.t. }\ & x_1 + x_2 + x_3 = 3 \\ & \alpha_1.x_1 + \alpha_2.x_2 + x_4 = 2 + \alpha_1 \\ & x_1 \leq 2,\ x_j \geq 0,\ j = 1,2,3,4 \end{aligned} \quad [4.6]$$

where the coefficients α_1 and α_2 are random, satisfying: $\alpha_1 = 1$ and $\alpha_2 = \frac{3}{4}$ with probability 0.25; or $\alpha_1 = -3$ and $\alpha_2 = \frac{5}{4}$ with probability 0.75. This means that there are two possible equations (or scenarios): either the second constraint is $x_1 + \frac{3}{4}x_2 + x_4 = 3$ (scenario occurs with probability 0.25), or it is $-3x_1 + \frac{5}{4}x_2 + x_4 = -1$ (scenario occurs with probability 0.75).

4.4.1. *Models with probability thresholds on the constraints*

In some cases, we wish to solve a linear problem *a priori* in such a way that the constraints are satisfied with some probability. This is known as chance constraint programming. The stochastic program is then expressed as follows:

$$\begin{aligned} \min\ & c^T.x \\ \text{s.t. }\ & A.x = b \\ & P(\beta.x = \gamma) > p \\ & x \geq 0 \end{aligned} \quad [4.7]$$

where β and γ are random variables and p is a fixed bound ($0 < p < 1$).

4.5. Stochastic linear programs with recourse

A two-step stochastic linear program with recourse can be written in the following general form:

$$\min z = c^T.x + \mathbb{E}_\xi[\min(q(\omega)^T.y(\omega))], \quad [4.8]$$

$$\text{s.t. } A.x = b, \quad [4.9]$$

$$T(\omega)x + W.y(\omega) = h(\omega), \quad [4.10]$$

$$x \geq 0, \quad y(\omega) \geq 0, \quad [4.11]$$

where $\mathbb{E}_\xi$ denotes the mathematical expectation taking into account the random variable ξ, $c \in \mathbb{R}^n$, $b \in \mathbb{R}^m$, $A \in \mathbb{R}^{m_1 \times n_1}$ and $W \in \mathbb{R}^{m_2 \times n_2}$.

The matrix W is called the (fixed) recourse matrix. It give us a convenient way to characterize the feasible region in the calculation:

$$\forall\omega,\ T(\omega) \in \mathbb{R}^{m_2 \times n_1},\ q(\omega) \in \mathbb{R}^{n_2},\ h(\omega) \in \mathbb{R}^{m_2},$$

Set: $\xi^T = (q(\omega)^T, h(\omega)^T, T_i(\omega))_{i=1,\dots,m_2} \in \mathbb{R}^N$, and $N = n_2 + m_2 + m_2 \times n_1$.

$T_i(\omega)$ is the ith component of the technology matrix.

$\Xi \subset \mathbb{R}^N$ is the support of ξ (the smallest closed subset of $\mathbb{R}^N$ such that $\mathbb{P}\{\xi \in \Xi\} = 1$).

The deterministic equivalent problem (DEP) is formulated as follows:

$$\begin{aligned} \min\ & z = c^T.x + \mathcal{Q}(x), \\ \text{s.t.}\ & A.x = b, \\ & x \geq 0, \end{aligned} \qquad [4.12]$$

with $\mathcal{Q}(x) = \mathbb{E}_\xi Q(x, \xi(\omega))$, and $Q(x, \xi(\omega)) = \min_y (q(\omega)^T.y | W.y(\omega) = h(\omega) - T(\omega)x, y \geq 0)$.

4.5.1. *L-shaped method*

In stochastic programs with recourse, the computations focus on two-step problems with finitely many realizations. The general formulation from equation [4.12] is used to construct a master problem as a function of x, and the recourse function is only evaluated as a subproblem. To do this, we assume that ξ is finitely supported with the realizations:

$$p_k = \mathbb{P}(\xi = k), k = 1, 2, \dots, K.$$

The DEP is expressed in extensive form (EF) as follows:

$$\begin{aligned} \min\ & c^T.x + \textstyle\sum_{k=1}^{k=K} p_k q_k^T y_k, \\ \text{s.t.}\ & A.x = b, \\ & T_k x + W.y_k = h_k,\ k = 1, 2, \dots, K \\ & x \geq 0, y_k \geq 0,\ k = 1, 2, \dots, K \end{aligned} \qquad [4.13]$$

The block of the EF is of the form:

$$\begin{pmatrix} A & . & . & \cdots \\ T_1 & W & . & \cdots \\ T_2 & . & W & \cdots \\ \vdots & \vdots & \ddots & \vdots \\ T_k & . & \cdots & W \end{pmatrix}.$$

The algorithm is presented in Algorithm 4.1.

Algorithm 4.1 L-shaped algorithm

Step 0: Set $r = s = \nu = 0$,
Step 1: Set $\nu = \nu + 1$,
Solve the following linear program:

$$\begin{aligned} \min\ & z = c^T.x + \theta, \\ \text{s.t.}\ & A.x = b, \\ & D_\ell.x \geq d_\ell, \ell = 1, 2, \ldots, r \\ & E_\ell.x + \theta \geq e_\ell, \ell = 1, 2, \ldots, s \\ & x \geq 0,\ \theta \in \mathbb{R} \end{aligned} \qquad [4.14]$$

Let (x^ν, θ^ν) be the optimal solution
If the constraint [4.14] is not present, then $\theta^\nu \leftarrow -\infty$
Step 2: $k = 1, 2, \ldots, K$, solve the linear program:

$$\begin{aligned} \min\ & \omega' = e^T.v^+ + e^T.v^-, \\ \text{s.t.}\ & W.y + I.v^+ - I.v^- = h_k - T_k.x^\nu, \\ & y \geq 0, v^+ \geq 0, v^- \geq 0 \end{aligned} \qquad [4.15]$$

with $e^T = \begin{pmatrix} 1 & 1 & \ldots & 1 \end{pmatrix}$
Let σ^ν be the associated multipliers, and define the operators:

$$D_{r+1} = (\sigma^\nu)^T.T_k \qquad [4.16]$$

$$d_{r+1} = (\sigma^\nu)^T.h_k \qquad [4.17]$$

To generate the constraint known as the feasibility cut, set:
$r \leftarrow r + 1$ added to the constraint [4.14] and return to **step (1):**
Step 3: $k = 1, 2, \ldots, K$, solve the linear program:

$$\begin{aligned} \min\ & \omega = q_k^T.y, \\ \text{s.t.}\ & W.y = h_k - T_k.x^\nu, \\ & y \geq 0, \end{aligned} \qquad [4.18]$$

Let Π_k^ν be the simple multipliers associated with the optimal solution of problem [4.18], and define the operators:

$$E_{s+1} = \Sigma_{k=1}^K p_k (\pi_k^\nu)^T T_k \qquad [4.19]$$
$$e_{s+1} = \Sigma_{k=1}^K p_k (\pi_k^\nu)^T h_k \qquad [4.20]$$

Set:

$$\omega^\nu = e_{s+1} - E_{s+1}.x^\nu \qquad [4.21]$$

$$\text{If} \qquad \theta^\nu \geq \omega^\nu \ \text{Stop.} \qquad [4.22]$$

x^ν is the optimal solution,
Else,
$s \leftarrow s + 1$ added to the constraint [4.14], and return to **step (1)**.

Algorithm 4.2 Multicut L-shaped algorithm

Step 0: Set $r = \nu = 0$, and $s_k = 0, \forall k = 1, \ldots, K$.
Step 1: Set $\nu = \nu + 1$,
Solve the following linear program:

$$\begin{aligned} \min\ & z = c^T.x + \textstyle\sum_{k=1}^K \theta_k, \\ \text{s.t.}\ & A.x = b, \\ & D_\ell.x \geq d_\ell, \ell = 1, 2, \ldots r \\ & E_{\ell(k)}.x + \theta_k \geq e_{\ell(k)}, \ell(k) = 1, 2, \ldots s(k) \\ & x \geq 0, \quad \forall k = 1, \ldots, K, \end{aligned} \quad [4.23]$$

Let $(x^\nu, \theta_1^\nu, \ldots, \theta_k^\nu)$ be the optimal solution
If the constraint [4.23] is not present, then $\theta_k^\nu \leftarrow -\infty \quad \forall k$, it will not be considered when calculating x^ν
Step 2: Same as previous algorithm.
Step 3: For all $k = 1, 2, \ldots, K$, solve the linear program:

$$\begin{aligned} \min\ & \omega = q_k^T.y, \\ \text{s.t.}\ & W.y = h_k - T_k.x^\nu, \\ & y \geq 0, \end{aligned} \quad [4.24]$$

Let Π_k^ν be the simple multipliers associated with the optimal solution of problem [4.24],
if

$$\theta_k^\nu < p_k(\Pi_k^\nu)^T.(h_k - T_k.x^\nu), \quad [4.25]$$

define the operators:

$$E_{s(k)+1} = p_k(\Pi_k^\nu)^T.T_k, \quad [4.26]$$

$$e_{s(k)+1} = p_k(\Pi_k^\nu)^T.h_k, \quad [4.27]$$

$$s(k) \leftarrow s(k) + 1, \quad [4.28]$$

If [4.25] is not satisfied $\forall k = 1, 2, \ldots K$, stop, x^ν is the optimal solution.
Else return to **step (1)**.

4.5.2. *Multicut L-shaped method*

At **step 3** of the L-shaped method, the K realizations of the program in the second step are optimized to find simple optimal multipliers after aggregating in [4.19] and [4.20] to generate the cuts [4.14].

4.5.3. *Interior linearization method*

This approach can be deduced from the L-shaped method by taking the dual linear program of [4.14]–[4.14]:

$$\begin{aligned} \max\ & \zeta = \rho^T.b + \textstyle\sum_{\ell=1}^{r} \sigma_\ell.d_\ell + \sum_{\ell=1}^{s} \pi_\ell.e_\ell \\ \text{s.t.}\ & \rho^T.A + \textstyle\sum_{\ell=1}^{r} \sigma_\ell.D_\ell + \sum_{\ell=1}^{s} \pi_\ell.E_\ell \leq c^T \\ & D_\ell.x \geq d_\ell, \quad \ell = 1, 2, \ldots, r \\ & \sigma_\ell \geq 0, \quad \textstyle\sum_{\ell=1}^{s} \pi_\ell = 1, \ell = 1, 2, \ldots, r \\ & \pi_\ell \geq 0, \quad \ell = 1, 2, \ldots, r \end{aligned} \qquad [4.29]$$

Now consider the dual of [4.24]:

$$\max\ \vartheta = \pi^T.(h_k - T_k.x^\nu), \qquad [4.30]$$

$$\text{s.t.} \quad \pi^T.W \leq q^T \qquad [4.31]$$

If the function [4.30] is not bounded for all k, we can find σ^ν satisfying:

$$(\sigma^\nu)^T.W \leq 0 \qquad [4.32]$$

$$\text{and} \quad (\sigma^\nu)^T.(h_k - T_k.x^\nu) > 0$$

or [4.15]–[4.17] admits a positive feasible dual solution (optimal primal solution).

Therefore, **step 2** of the L-shaped algorithm is equivalent to checking whether the function [4.30] is not bounded for every k. If so, define D_{r+1} and d_{r+1} as in [4.16]–[4.17] and add them to [4.30]–[4.31].

4.6. Nonlinear stochastic programming

4.6.1. *Approaches to two-step problems with recourse*

A two-step nonlinear stochastic program with recourse is defined as follows:

$$\min z = f^1(x) + \mathcal{Q}(x), \qquad [4.33]$$

$$g_i^1(x) \leq 0, \ i = 1, \ldots, \bar{m}_1 \qquad [4.34]$$

$$g_i^1(x) = 0, \ i = \bar{m}_1 + 1, \ldots, m_1 \qquad [4.35]$$

with $\mathcal{Q}(x) = \mathbb{E}_\omega[Q(x,\omega)]$, and

$$\min f^2(y(\omega),\omega) = Q(x,\omega) \qquad [4.36]$$

$$\text{s.c. } b_i^2(x,\omega) + g_i^2(y(\omega),\omega) \leq 0 \qquad [4.37]$$

$$f_i^2(x,\omega) + g_i^2(y(\omega),\omega) = 0 \qquad [4.38]$$

Algorithm 4.3 Interior linearization algorithm

Step 0: Set $r = s = \nu = 0$,
Step 1: Set $\nu = \nu + 1$,
Solve the linear program [4.29]–[4.31].
Let $(\rho^\nu, \sigma^\nu, \pi^\nu)$ be the optimal solution, and let (x^ν, θ^ν) be the dual solution.
Step 2: $\forall k = 1, 2, \ldots, K$, solve [4.30].
If the program [4.30] is not infeasible, stop.
If, $\forall k$, there exists an unbounded solution with outer radius σ^ν, construct the columns:

$$\begin{aligned} D_{r+1} &= (\sigma^\nu)^T.T_k, \\ d_{r+1} &= (\sigma^\nu)^T.h_k, \\ r &\leftarrow r+1 \end{aligned}$$

return to **step (1),**
if [4.30] is solvable, form the vectors E_{s+1} and e_{s+1} as in [4.19]–[4.20],
if $e_{s+1} - E_{s+1}.x^\nu - \theta^\nu \leq 0$, stop,
$(\rho^\nu, \sigma^\nu, \pi^\nu)$ and (x^ν, θ^ν) are optimal.
If $e_{s+1} - E_{s+1}.x^\nu - \theta^\nu > 0$, do $s \leftarrow s+1$ and return to **step (1)**.

4.6.2. *Regularized decomposition method*

This method uses the multicut approach to represent the function in the second step while incorporating a regularized quadratic term into the objective function.

4.6.3. *Methods based on the Lagrangian*

Consider a stochastic program as defined in [4.33]–[4.38]. The basic idea of this approach is to incorporate the first and second steps into the objective function in order to find the search direction. To do this, consider the following form:

$$\min z = f^1(x) + \mathbb{E}_\omega[f^2(y(\omega),\omega)], \qquad [4.39]$$

$$g_i^1(x) \leq 0, \ i = 1, \ldots, m_1 \qquad [4.40]$$

$$b_i^2(x,\omega) + g_i^2(y(\omega),\omega) \leq 0, \ i = 1, \ldots, m_2 \qquad [4.41]$$

Algorithm 4.4 Regularized decomposition algorithm

Step 0: Set $r = \nu = 0$, and $s(k) = 0, \forall k = 1, \ldots, K$.
select: a^1 feasible solution
Step 1: Set $\nu = \nu + 1$,
Solve the regularized master program:

$$\min c^T.x + \sum_{k=1}^{K} \theta_k + \frac{1}{2}\|x - a^\nu\|^2,$$

$$\text{s.t. } A.x = b,$$

$$D_\ell.x \geq d_\ell, \quad \ell = 1, 2, \ldots r$$

$$E_{\ell(k)}.x + \theta_k \geq e_{\ell(k)}, \quad \ell(k) = 1, 2, \ldots, s(k)$$

$$x \geq 0, \quad \forall k = 1, \ldots, K,$$

Let (x^ν, θ^ν) be the optimal solution, where $(\theta^\nu)^T = (\theta_1^\nu, \ldots \theta_K^\nu)$.
If $s(k) = 0$, then θ_k^ν can be neglected in the computation,
if $c^T x^\nu + e^T\theta^\nu = c^T a^\nu + \mathcal{Q}(a^\nu)$, stop a^ν is optimal.
Step 2: As in the previous algorithm, if the feasible cut [4.14] is generated, then $a^{\nu+1} \leftarrow a^\nu$ (infeasible step zero).
Return to **step (1)**.
Step 3: $\forall k = 1, 2, \ldots, K$, solve the linear subproblem [4.24], then compute $\mathcal{Q}_(x^\nu)$
if $\theta_k^\nu \prec p_k(\Pi_k^\nu)^T.(h_k - T_k.x^\nu)$,
add an optimality cut using the formulas [4.26]–[4.27].
Step 4: If [4.25] is not satisfied $\forall k = 1, \ldots, K$, then $a^{\nu+1} \leftarrow x^\nu$, return to **step (1).**
Step 5:
if $c^T.x^\nu + \mathcal{Q}_(x^\nu) \preceq c^T.a^\nu + \mathcal{Q}_(a^\nu)$,
then $a^{\nu+1} \leftarrow a^\nu$ return to **step (1),**
else $a^{\nu+1} \leftarrow a^\nu$, return to **step (1)**.

Let (λ, π) be the vector of multipliers associated with the constraints. The dual program of [4.39]–[4.41] is then formulated as follows:

$$\max \vartheta = \theta(\pi)$$

$$\pi(\omega) \succeq 0$$

$$\text{where } z = f^1(x) + \mathbb{E}_\omega[f^2(y(\omega), \omega)] \quad [4.42]$$

$$+ \mathbb{E}_\omega\left[\sum_{i=1}^{m_2} \pi(\omega)_i(b_i^2(x, \omega) + g_i^2(y(\omega), \omega))\right], \quad [4.43]$$

$$\text{s.t. } g_i^1(x) \leq 0, \quad i = 1, \ldots, m_1 \quad [4.44]$$

Algorithm 4.5 Algorithm based on the Lagrangian

Step 0: Set $\Pi^0 \geq 0$ and $\nu = 0$, go to **step (1).**
Step 1: Set $\Pi = \Pi^\nu$ in [4.43],
let $(x^\nu, y_1^\nu, \ldots, y_k^\nu)$ be the solution,
let $\hat{\Pi}_i^k = 0$, if $\hat{\Pi}_i^{\nu,k} = 0$, and $b_i^2(x^\nu, k) + g_i^2(y_k^\nu, k) \preceq 0$
$\hat{\Pi}_i^k = b_i^2(x^\nu, k) + g_i^2(y_k^\nu, k)$,
if $\hat{\Pi}_i^k = 0, \forall k$, stop.
Step 2 : Let:

$$\lambda^\nu = \min \theta(\pi^\nu, \lambda\hat{\pi}) \quad [4.45]$$

$$\text{s.t. } \pi^\nu + \lambda\hat{\pi} \succeq 0 \quad [4.46]$$

$$\lambda \succeq 0$$

let $\pi^{\nu+1} = \pi^\nu + \lambda^\nu \hat{\pi}$
set $\nu \leftarrow \nu + 1$, and go to **step (1).**

REMARK.– This algorithm either converges to the optimal solution in finitely many iterations or produces an infinite sequence that converges to the optimal point.

4.6.4. *Frank–Wolfe method for problems with simple recourse*

In the previous sections, we considered nonlinear problems that can be modeled by a deterministic equivalent with finitely many realizations of the random variable.

In the continuous case, problems with simple recourse can be expressed as follows:

$$\min c^T.x + \sum_{i=1}^{m_2} \Psi_i(\chi_i), \quad [4.47]$$

$$\text{s.t. } A.x = b, \quad [4.48]$$

$$T.x - \chi = 0, \; x \succeq 0, \quad [4.49]$$

where

$$\Psi_i(\chi_i) = \int_{h_i \leq \chi_i} q^-(\chi_i - h_i) dF_i(h_i) + \int_{h_i \succ \chi_i} q^+(h_i - \chi_i) dF(h_i) \quad [4.50]$$

but

$$\begin{aligned} Q_i(x) &= q_i^+ h_i^- - (q_i^+ - q_i F_i(T_i x)) T_i x \\ &\quad - q_i \int_{h_i \le T_i x} h_i dF_i(h_i) \end{aligned} \qquad [4.51]$$

which gives

$$\Psi_i(\chi_i) = q_i^+ h_i^- - (q_i^+ - q_i F_i(\chi_i))\chi_i - q_i \int_{h_i \preceq \chi_i} h_i dF_i(h_i).$$

The basic procedure of the Frank–Wolfe method is to approximate the objective function using the gradient and solve a linear program to find a direction of descent. Each $v.ah_i$ is assumed to admit an absolutely continuous distribution function F_i, and each Ψ_i is assumed to be differentiable. In this case, $\nabla\Psi(Tx) = (q^+ - q)^T(\bar{F}).T$, where $\bar{F} = diag\{F_i(T_i.x)\}$.

Algorithm 4.6 Frank–Wolfe algorithm

Step 0: Suppose that x^0 is a feasible solution of [4.47]–[4.49], set $\nu = 0$, and go to **step (1).**

Step 1: Solve the program:

$$\min z = (c^T + (q^+ - q)^T(\bar{F}^\nu))^T x, \qquad [4.52]$$

$$\text{s.t. } A.x = b, \qquad [4.53]$$

$$x \succeq 0,$$

where $\bar{F}^\nu = diag\{F_i(T_i.x^\nu)\}$.

Step 2: Find $x^{\nu+1}$ that minimizes:

$$c^T(x^\nu + \lambda(\hat{x}^\nu - x^\nu)) + \sum_{i=1}^{m_2} \Psi(T(x^\nu + \lambda(\hat{x}^\nu - x^\nu))), \quad \text{where} \quad 0 \le \lambda \le 1, \qquad [4.54]$$

If $x^{\nu+1} = x^\nu$, stop.

Else $\nu = \nu + 1$, and return to **step (1).**

4.6.5. *Approximation by sampling average: Monte Carlo method*

We start with the stochastic program:

$$\inf \quad \int_{\Xi} g(x,\xi)\mathbb{P}(d\xi) \qquad [4.55]$$

$x \in X$ and $X \subset \mathbb{R}^n$.

The random variable ξ is defined on the probability space $(\Xi, \mathfrak{B}, \mathbb{P})$.

Suppose that [4.55] admits an optimal solution x^*. The natural approach to solving [4.55] is to consider ν samples of ξ with the discrete distribution $\mathbb{P}^\nu$.

Let ξ^i be an i.i.d. sample of ξ. The problem [4.55] now becomes:

$$\inf \ (\tfrac{1}{\nu}) \sum_{i=1}^{\nu} g(x,\xi^i)\mathbb{P}(d\xi). \qquad [4.56]$$

Let x^ν be the random vector of solutions of [4.56]. The general idea is to find a distribution u such that:

$$\sqrt{\nu}(x^\nu - x^*) \xrightarrow[in\ distribution]{} u. \qquad [4.57]$$

The properties of u can be used to define confidence intervals for x^*.

THEOREM 4.1.– Suppose that $g(.,\xi)$ is convex and twice continuously differentiable, X is a convex polyhedron, and $\nabla g : \Xi \times \mathbb{R}^n \to \mathbb{R}^n$, satisfying:

– measurable: for every $x \in X$;

– Lipschitz: $\exists a \colon \Xi \to \mathbb{R}$, $\int_{\Xi} \|a(\xi)\|^2 \mathbb{P}(d\xi) < \infty$, such that:

$$|\ \nabla g(x_1,\xi) - g(x_2,\xi)\ | \leq a(\xi)\ |\ x_1 - x_2\ | \quad \forall x_1, x_2 \in X$$

$$\exists x \in X, \quad \int_{\Xi} \|g(x,\xi)\|^2 \mathbb{P}(d\xi) < \infty,$$

$$(x_1 - x_2)^T.G^*.(x_1 - x_2) > 0,\ \forall x_1, x_2 \in X,\ x_1 \neq x_2$$

where $G^* = \int_{\Xi} \nabla^2 g(x^*,\xi)\mathbb{P}(d\xi)$.

Then the solution x^ν of [4.56] satisfies:

$$\sqrt{\nu}(x^\nu - x^*) \xrightarrow[in\ distribution]{} u, \qquad [4.58]$$

where u is the solution of:

$$\begin{array}{ll} \min & \frac{1}{2}u^T.G.u + c^T.u \\ \text{s.t.} & A_i.u_i \leq 0, \ i \in I(x^*) \\ & u^T \nabla \bar{g}^* = 0, \end{array} \qquad [4.59]$$

$$\begin{array}{l} X \leftarrow \{x \mid Ax \leq b\}, \ \ I(x^*) \leftarrow \{i \mid A_i x^* = b\}, \ \ c \sim N(0, \Sigma^*), \\ \Sigma^* \leftarrow \int_\Xi (\nabla g(x^*, \xi) - \nabla \bar{g}^*)(\nabla g(x^*, \xi) - \nabla \bar{g}^*)^T \mathbb{P}(d\xi). \end{array}$$

4.6.6. *Stochastic gradient method*

Consider a probability space $(\Omega, \mathcal{A}, \mathbb{P})$ and a random variable ω defined on Ω taking values in a probability space $(\mathbb{W}, \mathfrak{W})$. Write μ for the probability law obtained by transporting the law $\mathbb{P}$ under ω. Let U be a Hilbert space (whose scalar product and norm are denoted $\prec .,. \succ$ and $\|.\|$), U^{ad} a closed non-empty convex subset of $\mathbb{U}$ and j a function defined on $\mathbb{U} \times \mathbb{W}$ taking values in $\bar{\mathbb{R}}$. Write J for the expected value of the function j (assumed to be integrable for every $u \in U^{ad}$):

$$J(u) = \mathbb{E}(J(u, \omega)) \qquad [4.60]$$

$$= \int_\Omega j(u, \omega(\omega)) d\mathbb{P}(\omega) \qquad [4.61]$$

$$= \int_{\mathbb{W}} j(u, \omega) d\mu(\omega),$$

and consider the problem: $\min_{u \in U^{ad}} J(u)$.

With the classical hypotheses of convexity and differentiability, provided that we are prepared to compute the gradient $J^{'}(u)$ of J at every point u, we can use a gradient-type algorithm (optimal step, conjugate gradient, quasi-Newton, Newton, etc.) to solve the problem [4.60]–[4.61]. This approach can, however, be extremely expensive in terms of computation time, since each evaluation of the gradient requires us to compute the expected value over the space $\mathbb{W}$, which can potentially be very large. In fact, the probabilistic structure of the problem is not strictly used in the optimization, since it is fully taken into account when the gradient is computed.

Algorithm 4.7 Stochastic gradient algorithm

Step 0: Choose an initial value $u^{(0)} \in U^{ad}$, and a sequence of positive real numbers $\{\epsilon^k\}_{k\in\mathbb{N}}$.
Step 1: At the kth iteration of the algorithm, fix a realization ω^{k+1} of the r.v. ω in accordance with its law, independently of the realizations $(\omega^1,\dots,\omega^k)$ from previous iterations.
Step 2: Compute the gradient of j with respect to u at the point (u^k,ω^{k+1}) and update u as follows:

$$u^{k+1} \leftarrow proj_{U^{ad}}\{u^{(k)} - \epsilon^{(k)} j'_u((u^{(k)},\omega^{k+1}))\}.$$

Step 3: $k \leftarrow k+1$,
return to **step (2)**.

4.7. Stochastic dynamic programming

Stochastic dynamic programming uses the tools of dynamic programming while accounting for non-deterministic (probabilistic) actions.

The state at time $t+1$ depends on:

– the state at time t;

– the action chosen at time t;

– the probability distribution.

A dynamic program can be expressed as follows:

$$x_{k+1} = f_k(x_k, u_k, \omega_k), \qquad k = 0, 1, \dots, N-1 \qquad [4.62]$$

– x_k: the state of the system at time k (the initial state is given);

– $u_k = \mu_k(x_k)$: the decision function;

– ω_k: random parameter, also called the perturbation;

– N: the problem horizon (number of decisions applied);

– f_k: the transfer function at step k.

For stochastic systems, the functions f_k depend on the random perturbations ω_k and are therefore also random variables. At this point, it is impossible to minimize the total costs. The traditional approach is to minimize their expected values.

4.7.1. *Markov decision process*

A Markov decision process (MDP) is a stochastic process that represents the evolution of a dynamic system controlled by sequences of actions (decisions). The objective of this type of problem is to make an optimal selection of actions to control the system. This is also known as stochastic optimal control.

The Markov decision process is the set $\{\mathbf{S}, \mathbf{A}_s, \mathbb{P}_{st}(a), \mathbf{C}(s,a), V_\alpha\}$, where:

– $\mathbf{S}$: countable set of observed states, $\mathbf{S} \subset \mathbb{N}^d,\ d \in \mathbb{N}^*$;

– $\mathbf{A_s}$: set of admissible actions in the state s, setting: $\mathbf{A} = \bigcup_{s\in\mathbf{S}} \mathbf{A_s}$;

– $\mathbb{P}_{st}(a)$: transition probability from the state s to the state $\mathbf{t}$ when making the decision $\mathbf{a}$;

– for each step n, we define a reward function (in other words, an instantaneous cost function) such that:

$$\begin{aligned} \mathbf{C}\colon \mathbf{S} \times \mathbf{A} &\to \mathbb{R} \\ (s,a) &\mapsto C(s,a) \end{aligned} \tag{4.63}$$

This represents the cost incurred in the state s when making the decision $\mathbf{a}$ at time n and arriving at the state s$'$;

– $\mathbf{V}_\alpha$: updated cost criterion of factor $\alpha \in]0,1[$.

Below, we present two algorithms for solving MDP problems: the value iteration algorithm with output $\mu(x)\ \forall x \in S$ and the policy iteration algorithm.

The policy iteration algorithm is a very powerful technique for computing the optimal policy instead of finding it indirectly through the intermediary of the optimal value function. It alternates between phases of evaluating and improving the policy. The evaluation phase evaluates the current policy π with the value iteration algorithm. The improvement phase then transforms the policy π into another policy π' such that:

$$V_{\pi'}(x) \geq V_\pi(x) \qquad \forall x \in S.$$

The value function of the policy π is computed by solving the system of equations:

$$V_\pi^\alpha(x) = C(x, \pi(x)) + \alpha \sum_{y\in\mathbb{N}^2} P_{xy}(a) V_\pi^\alpha(y). \tag{4.64}$$

Algorithm 4.8 Value iteration algorithm with $\mu(x)$ $\forall x \in S$

Initialization: $V(x) \leftarrow 0$;
$\delta \leftarrow 0$; Repeat:
$v \leftarrow V(x)$; For $x \in S$, do:
For $a \in A$, do:

$$Q(x,a) \leftarrow C(x,a) + \alpha \sum_{y \in \mathbb{N}^2} P_{xy}(a) V_n^\alpha(y)$$

end for

$$V(x) = \min_{a \in A} Q(x,a)$$

end for
$\delta \leftarrow \arg\min(\delta, |v - V(x)|)$
until $\delta < \epsilon$
for $x \in S$, do
$\mu(x) \leftarrow \arg\min_{a \in A} Q(x,a)$
end for

Now, the policy of each state is improved:

$$\pi'(x) = \arg \min_{a \in A(x)} (C(x, \pi(x)) + \alpha \sum_{y \in \mathbb{N}^2} P_{xy}(a) V_n^\alpha y)) \qquad [4.65]$$

until $\pi \leftarrow \pi'$.

4.7.2. *Scenario tree*

When discretizing a stochastic dynamic optimization problem, we need to be careful to remember that we are faced with two probabilistic objects of different natures that can be handled separately [BAR 04]:

1) the expected value in the criterion, which is generally estimated by Monte Carlo methods;

2) measurability constraints.

In the case of multiple time steps and perfect memory, the "scenario tree" methodology proposes discretizing both of these objects simultaneously using chronicles organized into trees of scenarios. Let us present their implementation informally using Figure 4.1 as an illustration.

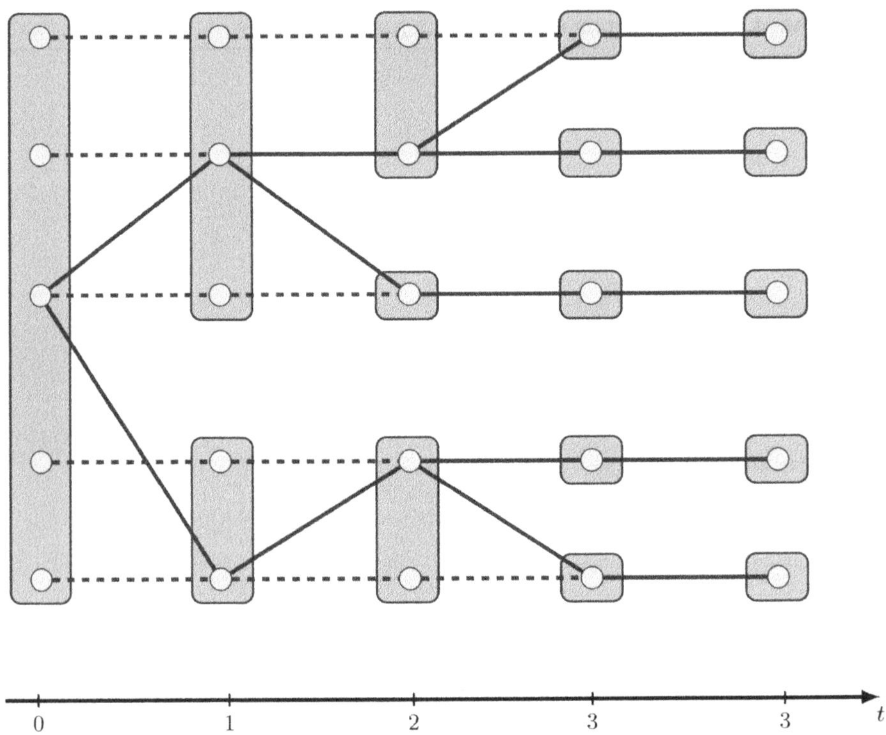

Figure 4.1. *Constructing a scenario tree. For a color version of this figure, see www.iste.co.uk/radi/optimizations.zip*

Consider a set of samples of the random variable $\mathbf{V}$. Formally, this is a set of independent and identically distributed (i.i.d.) random variables $\mathbf{V}^1, \ldots, \mathbf{V}^N$ with the same law as $\mathbf{V}$. In Figure 4.1, we chose $N = 5$ and a horizon of 5 time steps. The processes are symbolized by small circles connected with dotted lines. Choosing a sample of size N corresponds to performing a Monte Carlo–type estimate.

To represent the non-anticipativity constraint, these scenarios are grouped by shared history using an ad hoc metric [HEI 06]. The gray rectangles are the classes. Next, we select a representative from each class from the various noise samples at this time. We now have a tree structure that is represented by the bold lines such that the representative of each class is the point through which the line passes. Each node of the tree corresponds to a unique history; a stock variable and a control variable are defined at each node. Now, we rewrite the set of constraints on this tree, as well as the objective function. The problem is now formulated as a mathematical programming problem.

Finally, we need to bear in mind that scenario trees are a stochastic solving method in the sense that they yield a solution that depends on the realizations of the scenarios chosen when constructing the tree. This makes evaluating the quality of the solution more delicate (a large number of independent experiments is often required to evaluate the error).

The basic idea is as follows: represent the diffusion of information as a tree structure, as illustrated in Figure 4.2. Suppose that we are at $t = 0$ and that the realization w_0 of $\mathbf{W}_0$ is known; accordingly, our tree has just one root (node 0). Now, consider three possibilities for the future realizations of the random element $\mathbf{W}_1$, represented by three successors of the original node of the tree (nodes 1, 2 and 3). For each of these nodes, fix a realization of the random variable $\mathbf{W}_1$. This construction procedure is continued until the end of the problem horizon is reached.

Write $\mathcal{N}$ for the set of nodes, $\mathcal{R}$ for the set of roots and $\mathcal{L}$ for the set of leaves of the tree. After constructing this tree, we have effectively defined the following functions:

– the time function $\theta\colon \mathcal{N} \rightarrow \{0, 1, \ldots, T\}$ that sends each node to the corresponding time step; this also defines the multifunction θ^{-1} that sends each time step $t \in \{0, \ldots, T\}$ to the set of nodes of level t;

– the parent function $\pi\colon \mathcal{N} \rightarrow [0, 1]$ that sends each node of the tree to the node before it;

– the weight function $\pi\colon \mathcal{N} \rightarrow [0, 1]$ that sends each node of the tree to the probability of passing through this node (thus, the sum of the $\pi(i)$ for $i \in \theta^{-1}(t)$ is equal to 1 for every t);

– the multifunction $F = \nu^{-1}$ that sends each node i of the tree to the set of its children (with the convention that, if i is a terminal node of the tree, then $F(i) = \emptyset$);

– the function F^* that sends each node i to the subtree hanging from i, namely $F^*(i) = F(i) \cup F^2(i) \cup \cdots \cup F^T(i)$.

In practice, we often have a set of scenarios that are assumed to be independent and we wish to construct an optimal tree structure based on these scenarios in a sense that remains to be defined. There is a wealth of literature on constructing scenario trees [PFL 01, HEI 03] and on the stability of the resulting optimization problem with respect to this discretization [HEI 06].

4.8. Application to the reliability of mechanical systems

The reliability of mechanical and mechatronic systems is the foundation of the competitive strength of companies in France and across the globe in the automotive, aeronautic, aerospace, civil engineering and defense sectors, among others For example, in the field of mechatronic systems, namely, systems that combine mechanical and electronic elements, new failure phenomena have been encountered that have not been studied in depth and which are therefore not sufficiently understood.

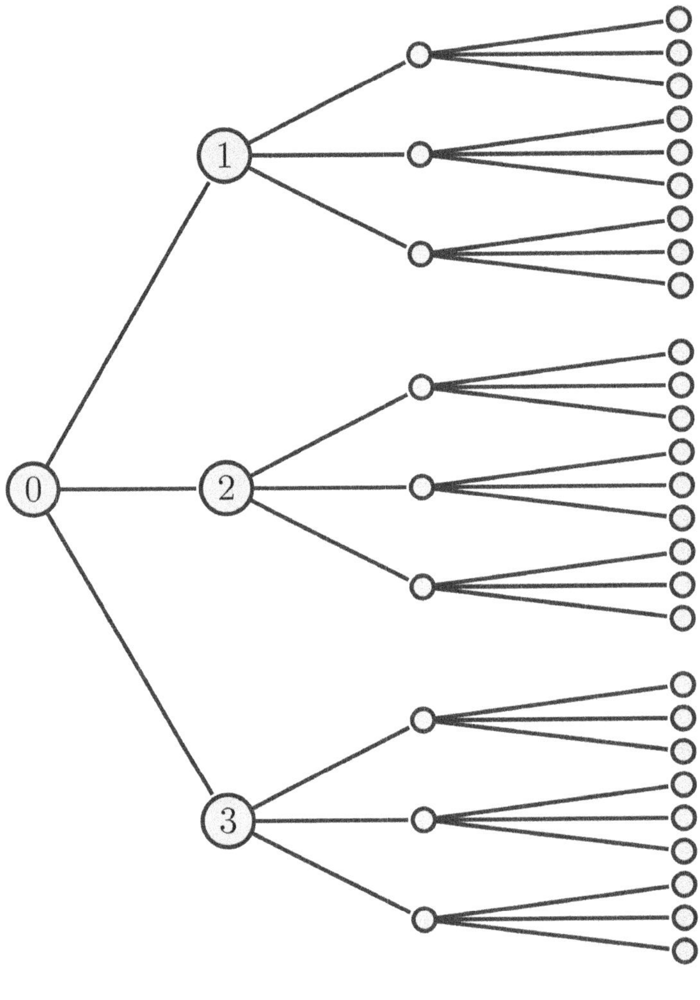

Figure 4.2. *Example of a tree structure. For a color version of this figure, see www.iste.co.uk/radi/optimizations.zip*

Especially in aeronautic and automotive applications, the reliability of embedded systems depends on several parameters. In the case of printed electronic circuits, excessive stress can cause failures such as welding or terminal fractures, or loss of signal and information. This can significantly impair system functionality and safety.

More than ever before, system reliability is therefore a major concern for industrial companies, which must comply with ever stronger demands from their customers. Failure may incur penalties or warranty extensions, and poor quality can be associated with significant costs to repair defects, or possibly even customer dissatisfaction or a bad reputation that compromises future markets. At the same time, embedded electronics are becoming ever more important in vehicles and systems.

A key factor of system reliability is being able to implement a rigorous monitoring framework that is capable of predicting and detecting the failure modes of the system in the development phase to avoid any major impact on the behavior and safety of the system. Such a monitoring framework can combine techniques of diagnosis and fault detection to guarantee optimal performance for processes and systems.

4.8.1. *Position and modeling of the reliability problem*

Structures are made to respond to a set of clearly identified needs. The goal of structure reliability is to evaluate the probability that a structure subject to random factors (vibrations, temperatures, shocks, fatigue, etc.) is capable of fully satisfying these needs over a certain lifetime. To calculate this probability, a rigorous four-step study methodology is required:

1) define a deterministic mechanical model suitable for the problem at hand;

2) identify the random parameters of this model and model them with a suitable probabilistic tool;

3) define the failure modes and scenarios of the problem;

4) evaluate the probability that these failure modes will occur.

The precision and accuracy of the result and the failure probability of the structure are therefore conditioned by a set of expert, experimental, mathematical, numerical and practical factors at each step of the reliability study.

4.8.1.1. *Deterministic mechanical model*

The first step of a reliability study is to identify the mechanical problem. A suitable deterministic mechanical model is then defined. This model incorporates the geometry of the structure, the mechanical properties of the materials and the boundary conditions.

4.8.1.2. *Randomness and probabilistic modeling*

Of the various parameters involved in the model, some are only known randomly. Randomness in the physical and material properties and geometric characteristics of the parts of the structure is called internal randomness. These uncertainties depend on the quality and precision of the manufacturing and characterization of the materials and structural elements. Randomness in the actions performed on the structure is called external randomness. The origin of these uncertainties often originates from the difficulties associated with precisely evaluating the value of a poorly understood stress. Examples include the action of waves, the wind or road traffic on a bridge.

The parameters are identified and suitable statistical tests are performed to validate or reject the hypotheses. If the problem has m random variables X_i, we define the random vector:

$$X = (X_1, X_2, \dots, X_m)^T. \quad [4.66]$$

4.8.1.3. *Failure modes of a structure*

To guarantee the integrity of the structure against the anticipated risks, rules are established to describe its proper working order. Violating one of these criteria triggers an elementary failure mode of the structure. The sequence of events leading to a failure is called a failure scenario. The failure modes or scenarios are formulated as a function of the potential risks: the presence of excessive stresses or strains in a structural element, the breakage of a part, etc.

Failure can be triggered by various physical phenomena: plastification, fatigue, creep, the presence of fissures, large deformations, etc. Each of these phenomena or combinations of them create a failure mode or scenario. To simplify the notation, we will consider the case of a structure with a single failure mode below. The probabilistic model of a failure mode is established by defining a function $G(X)$, called the limit state function or performance function. Note that the function $G(X)$ is itself a random variable by construction. Thus:

- $G(X) > 0$ defines the safe region of the structure;
- $G(X) < 0$ defines the failure region of the structure;
- $G(X) = 0$ defines the limit state surface.

The structure therefore has two possible states: a state corresponding to proper working order, and a failure state, separated by a so-called limit state boundary. Nevertheless, reality is often much more complex than this binary model.

4.8.1.4. *Failure probability of a structure*

DEFINITION 4.2.– *The failure probability of the structure is now equal to:*

$$P_f = P(G(X) \leq 0), \quad [4.67]$$

namely:

$$P_f = \int_{G(X)\leq 0} f_X(x)dx. \quad [4.68]$$

The term f_X denotes the joint probability density function of the random variable X.

DEFINITION 4.3.– *The reliability of the structure is defined as:*

$$P_s = 1 - P_f. \quad [4.69]$$

Although this formulation of P_f is straightforward, it can prove extremely complex to calculate. In general, the difficulties associated with explicitly determining f_X and the often nonlinear expression of the boundary of D_f make analytic computations impossible.

There are currently two major categories of method for evaluating the failure probability of a structure. The first aims to evaluate P_f on the entire failure region D_f using simulations based on the Monte Carlo technique. The second idealizes the region D_f to calculate an approximate value of P_f based on a reliability index β. The Hasofer and Lind index is presented below.

4.8.1.5. *Hasofer and Lind index*

To eliminate the risk of the reliability index depending on the limit state, Hasofer and Lind suggested calculating β in the space of standardized normally distributed and statistically independent random variables. To do this, the random vector X is transformed into a random vector U by:

$$U_i = T(X_i). \quad [4.70]$$

The random variables follow a standardized normal distribution, and, $\forall i \neq j$, the random variables U_i and U_j are mutually independent. This probabilistic transformation T requires the statistical distributions of each random variable to be known. After the transformation, the limit state function becomes:

$$H(U) = G[X(U)]. \quad [4.71]$$

The failure probability is now equal to:

$$P_f = P(H(U) \leq 0), \quad [4.72]$$

namely:

$$P_f = \int_{H(u)\leq 0} \Phi_m(u)du. \quad [4.73]$$

The term Φ_m is the density function of the standardized multinormal distribution. The reliability index β_{HL} is defined as the Euclidean distance from the origin of the standardized normal space to the limit state surface $H(u) = 0$ (Figure 4.3). The term

u denotes a realization of the random vector U, i.e. $u = (u_1, u_2, \ldots, u_m)^T$, and $H(u)$ denotes a realization of the random variable $H(U)$. We must therefore solve the following constrained stochastic minimization problem:

$$\begin{cases} \beta_{HL} = \min \sqrt{u^T.u} \\ u \in \mathbb{R}^m, \text{ satisfying:} \\ H(u) = 0 \end{cases} \quad [4.74]$$

This problem is equivalent to

$$\begin{cases} \beta_{HL} = \min f(u) \\ u \in \mathbb{R}^m, \text{ satisfying:} \\ H(u) = 0 \end{cases} \quad [4.75]$$

where

$$f(u) = \frac{1}{2} u^T.u. \quad [4.76]$$

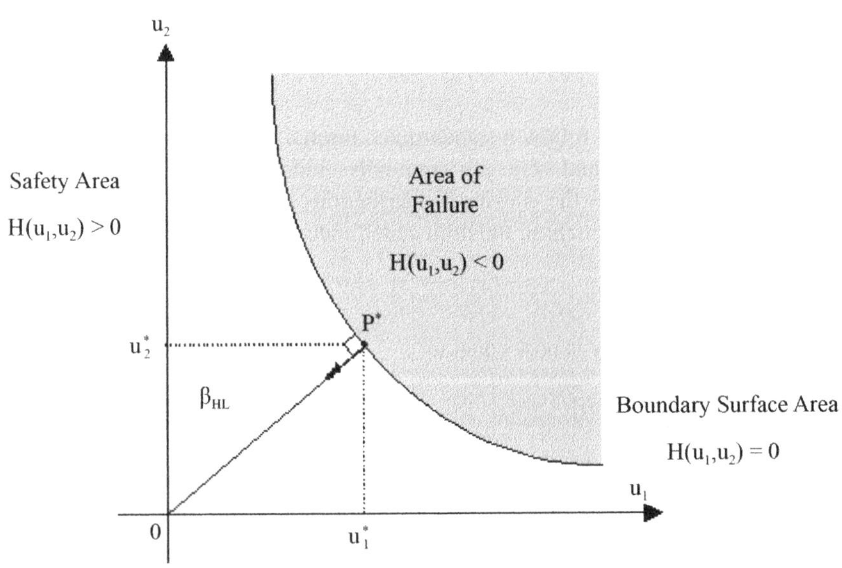

Figure 4.3. *Geometric representation of β_{HL} for a problem with two random variables*

The reliability index is therefore the minimum of the function f subject to the constraint $H(u) = 0$. Calculating β_{HL} analytically is only feasible in a limited number of special cases and a numerical solution is necessary in general. The objective function f is a quadratic, convex form that is twice continuously differentiable with respect to the variables u_i. However, the constraint function $H(u)$ is rarely convex. Furthermore, the function $H(u)$ is sometimes complicated, implicit and may fail to be continuously differentiable with respect to the variables u_i.

Let u^* be the vector solution of the problem [4.75], and let P^* be the point on the limit state surface such that $OP^* = u^*$. The point P^* is called the design point or most probable failure point (see Figure 4.4). Moreover:

$$u^* = -\beta_{HL}\alpha. \qquad [4.77]$$

The vector α is the normal to the surface $H(u) = 0$ at the point u^*, i.e. the normalized gradient at this point. Our goal is to relate β_{HL} to the failure probability of the structure. There are several methods for doing this that give values of P_f with varying resolutions.

4.8.1.6. *FORM method*

The FORM method (First-Order Reliability Method) replaces the limit state surface by the tangent hyperplane at the design point P^*, $Z(u) = 0$, which gives a linear approximation of $H(u) = 0$ (Figure 4.5). The general principle of the FORM method is as follows:

– *Step 1*: Define the probabilistic transformation T between the initial physical space and the standardized normal space.

– *Step 2*: Transform the failure surface $G(u) = 0$ in the initial space into the failure surface $H(u) = 0$ in the standardized normal space.

– *Step 3*: Compute the reliability index β_{HL} that solves the problem [4.75].

– *Step 4*: Compute P_f^* as the approximation of P_f defined by:

$$P_f^* = \int_{Z(u)\leq 0} \Phi_m(u)du, \qquad [4.78]$$

namely:

$$P_f^* = \Phi(-\beta_{HL}). \qquad [4.79]$$

The function Φ is the distribution function of the standardized normal law, and D_f^* is the region of $\mathbb{R}^m$ such that $Z(u) \leq 0$, where:

$$Z(u) = \alpha^T.u + \beta_{HL}. \qquad [4.80]$$

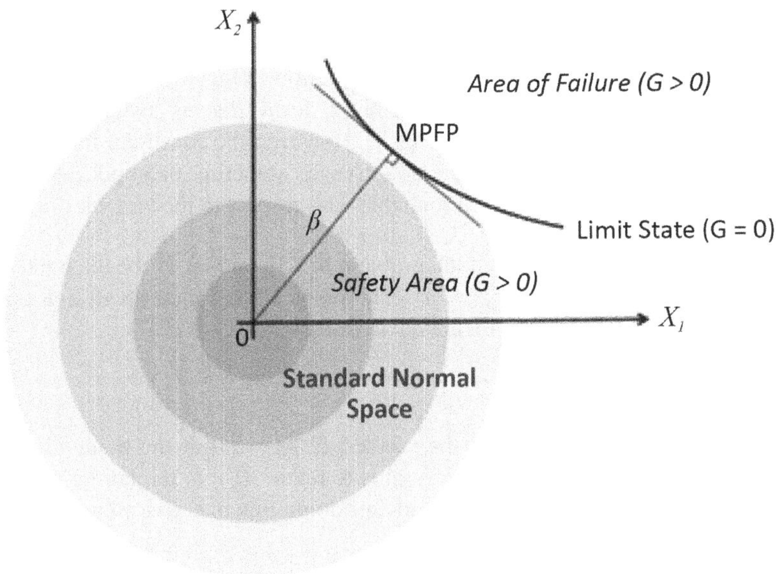

Figure 4.4. *Most probable failure point. For a color version of this figure, see www.iste.co.uk/radi/optimizations.zip*

If the limit state surface is:

– convex, then $P_f^* > P_f$;

– concave, then $P_f^* < P_f$;

– a hyperplane, then $P_f^* = P_f$.

The precision of the FORM approximation strongly depends on the curvature of the limit state surface at the design point and therefore on the nonlinearity of the function $H(u) = 0$. One of the major inconveniences of the FORM method is the difficulty of estimating the computational error in the failure probability. One approach to validating its results is to employ the Monte Carlo simulations presented above. The number of computations required depends directly on the number of random variables of the problem.

The FORM method yields a particularly rich set of results. Thus, we obtain an estimate of P_f, the most probable failure point P^* (MPFP) in the normed space, and, potentially, the sensitivity of the failure probability to the various random elements of the problem. Indeed, using equation [4.77], the directional cosines are defined by:

$$\alpha_i = -\frac{1}{\beta_{HL}} u_i^*. \qquad [4.81]$$

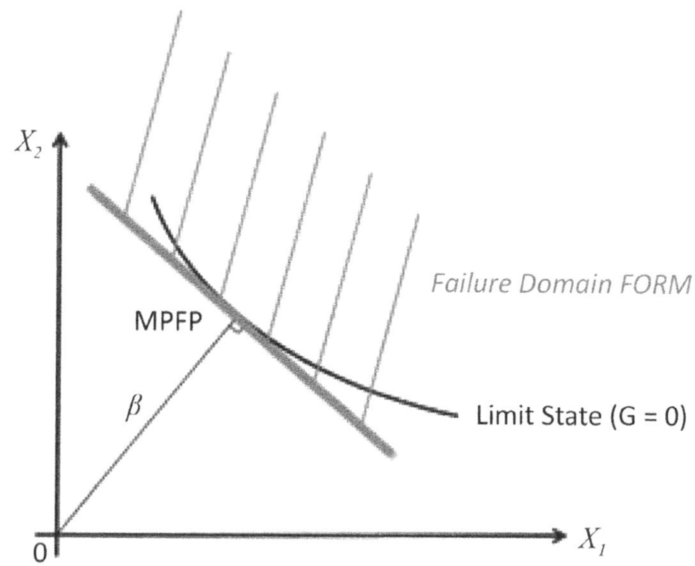

Figure 4.5. *Illustration of FORM. For a color version of this figure, see www.iste.co.uk/radi/optimizations.zip*

These directional cosines measure the importance of the random variables U_i in the objective function $\sqrt{u^T.u}$ at the most probable failure point u^*. A directional cosine of $\alpha_i \approx 0$ means that $u_i^* \approx 0$ and $x_i^* \approx E[X_i]$. The importance of u_i^* in the value of β_{HL} is therefore close to zero. But $P_f = \Phi(-\beta_{HL})$, so the statistical distribution of the variable X_i does not influence the value of the failure probability, and this parameter can be viewed as deterministic. Conversely, a directional cosine α_i with a high absolute value characterizes the stochastic importance of the random variable U_i in the failure probability of the structure. The parameters α_i^2 are called the importance factors of the random variables U_i. Moreover, under the probabilistic transformation T that relates U with X, they reflect the importance of the random variables in physical space if these variables are mutually independent.

EXAMPLE 4.2.– Consider a beam of reinforced concrete with a rectangular cross-section on simple supports subjected to moments M applied to its endpoints, as illustrated in Figure 4.6. For this beam, we define the following limit state function:

$$G(A_S, f_y, d, \eta, b, f_c', M) = M_R - M, \quad [4.82]$$

where M_R is the moment of resistance of the beam, given by

$$M_R = A_S f_y d \left[1 - \eta \left(\frac{A_S f_y}{b d f_c'} \right) \right], \quad [4.83]$$

and A_S is the surface of the cross-section, f_y is the elastic limit, d is the maximum distance between the points of the cross-section and the neutral plane, η is the stress parameter of the concrete, f_c' is the ultimate compressive stress of the concrete and b is the width of the cross-section.

Figure 4.6. *Diagram of the beam on simple supports of reinforced concrete*

The example considers different combinations of probability distribution types for the random variables, as shown in Table 4.1.

Variables	Mean	Coeff. of variation	Case A	Case B	Case C	Case D
A_S (cm^2)	10	0.225	Normal	Normal	Log-nor.	Log-nor.
f_y (MPa)	328.8	1.03	Normal	Normal	Log-nor.	Log-nor.
f_c' (MPa)	24.13	1.44	Normal	Normal	Log-nor.	Log-nor.
b (cm)	20.3	0.1	Normal	Normal	Log-nor.	Log-nor.
d (cm)	33.5	0.2	Normal	Normal	Log-nor.	Log-nor.
h (cm)	1.5	0.1	Normal	Normal	Log-nor.	Log-nor.
M (kNm)	36.2	0.019	Normal	Log-nor.	Normal	Log-nor.

Table 4.1. *Characteristics of the random variables for the concrete beam*

We will now use the FORM method to compute the reliability index. The results of the computations for estimating the reliability indices and failure probabilities are shown in Table 4.2 for each case.

	Values of the variables at the design points			
Variables	Case A	Case B	Case C	Case D
A_S (cm^2)	9.6	9.6	9.5	9.56
f_y (MPa)	170.3	186.8	213.9	232.2
f'_c (MPa)	23.09	22.9	21.2	21.6
b (cm)	20.3	20.3	20.2	20.2
d (cm)	29.9	29.9	28.0	29.0
h (cm)	1.5	1.5	1.5	1.5
M (kNm)	46.1	50.2	51.9	58.3
β-FORM	3.83	3.76	4.39	4.09
P_f (%)-FORM	$6.33 \cdot 10^{-5}$	$8.45 \cdot 10^{-5}$	$5.73 \cdot 10^{-6}$	$2.15 \cdot 10^{-5}$

Table 4.2. *Results obtained with the FORM method for the concrete beam*

The results show the sensitivity of the values of the failure probability with respect to the probability distributions chosen for the random variables. The Matlab script used to implement the FORM method is presented in the Matlab section.

For more details, the reader is invited to refer to [ELH 11].

4.9. Using Matlab

The Matlab script used to compute the reliability index using the FORM method is as follows:

```
clear all; clc;
delete *.asv; close all
tic format short g
x1 = 1.56;       dx1 = 0.036; x2 = 47.7;       dx2 = 0.15;
x3 = 3.5; dx3 = 0.21; x4 = 8;         dx4 = 0.045;
x5 = 13.2; dx5 = 0.086; x6 = 0.59;       dx6 = 0.05;
x7 = 326.25; dx7 = 0.17;
xmean    = [  x1 x2 x3 x4 x5 x6 x7 ].'; dist1   = 'Normal';
dist2  = 'Lognormal';
designVariable.x0    = xmean;
designVariable.Dist   = {dist2, [x1  x1*dx1 0]
                         dist2, [x2  x2*dx2 0]
```

```
                              dist2, [x3  x3*dx3 0]
                              dist2, [x4  x4*dx4 0]
                              dist2, [x5  x5*dx5 0]
                              dist2, [x6  x6*dx6 0]
                              dist2,[x7  x7*dx7 0]};

gFN = @mygFN;

GTol = feval(gFN,designVariable.x0)/1e5;

options = simpleFORMoptset('GTol',GTol,...
                           'Display','diagnose');

[x,beta,Pf,state] = simpleFORM(gFN,designVariable,options); beta

Pf = Pf*100 Rlevel = 100 - Pf
toc
```

For a color version of this code, see www.iste.co.uk/radi/optimizations.zip

with the following functions:

```
function [G] = mygFN(x)
G = x(1).*x(2).*x(5).*(1 -
(x(6).*x(1).*x(2))./(x(4).*x(5).*x(3))) - x(7);
return
```

For a color version of this code, see www.iste.co.uk/radi/optimizations.zip

and

```
function [x,beta,Pf,state] =
simpleFORM(gFN,designVariable,options,gFNOptions)

state = [];
designVariable = simpleRTChange2Normal(designVariable);

NbVariables = length(designVariable.x0);

% step 2
x                = designVariable.x0; distributions   =
designVariable.Dist(:,1); distParameters  =
designVariable.Dist(:,2);

distParametersN = []; distributionsN  = []; for counter = 1 :
length(x)
    distributionsN{counter} = 'Normal';
    distParametersN{counter} = [0 1];
```

```
end

msg = nargchk(1,4,nargin); if ¬isempty(msg)
    error('simpleFORM:numberOfInputs',msg);
end

if nargin≥3
    if ¬isempty(options) && ¬isa(options,'struct')
        error('simpleFORM:thirdInputNotStruct','Invalid Input for
            simpleFORM.');
    end

    if isempty(options)
        options = simpleFORMoptset;
    end

    if nargin == 3
        gFNOptions = [];
    end
end

if nargin==2
    options = simpleFORMoptset;
    gFNOptions = [];
end

if isempty(gFNOptions)
    getGradient = 'simpleRTGradient(gFN,x,NbVariables)';
    getG = 'feval(gFN,x)';
else
    getGradient = 'simpleRTGradient(gFN,x,NbVariables,gFNOptions)';
    getG = 'feval(gFN,x,gFNOptions)';
end
beta     = options.beta;
GTol     = options.GTol;

BetaTol = options.BetaTol;

%Create initial state
state = makeState(NbVariables,gFN,options); state.g0 = eval(getG);
state.gFNEval = state.gFNEval  + 1;

%%%%%%%%%%%%%%%%%%%%%%%%%%%%%%%%%%%%%%%%%%%%%%%%%%%%%%%%%%%%%%%%%%%%%%

if strcmpi(options.Display,'diagnose')
    simpleFORMdiagnose(gFN,NbVariables,options);
end

if  any(strcmpi(options.Display, {'iter','diagnose'}))
    fprintf('\n\n');
    fprintf('Iteration  g—count g(x) beta DeltaG  DeltaBeta\n');
end

%%%%%%%%%%%%%%%%%%%%%%%%%%%%%%%%%%%%%%%%%%%%%%%%%%%%%%%%%%%%%%%%%%%%%%
```

```
exitFlag = []; while isempty(exitFlag)
    state.Iteration = state.Iteration + 1;
    iteration = state.Iteration;
    stopCondition02 = 0;
    while stopCondition02 == 0
    % step 3
    fx = simpleRTGetPDF(x,distributions,distParameters);
    for counter = 1 : length(fx)
      if fx(counter) == 0
          fx(counter) = eps;
      end
      switch upper(distributions{counter})
%------------ ------------
      case {'NORMAL','NORM'}
        xstar(counter,1) = (x(counter)
            -distParameters{counter}(1))/distParameters{counter}(2);
      case {'LOGISTIC','LOG'}
        xstar(counter,1) = 1/(1 + exp( - (pi/sqrt(3))*((x(counter)
         -distParameters{counter}(1))/distParameters{counter}(2))));
      case {'LOGNORMAL','LOGN'}
        xstar(counter,1) = (log(x(counter))
         -distParameters{counter}(1))/distParameters{counter}(2);
        end
     end
        PHIinv = xstar;
        phi = simpleRTGetPDF(PHIinv,distributionsN,distParametersN);
        XstdN  = phi./fx;
        XmeanN = x - PHIinv.*XstdN;
        for counter = 1 : length(XstdN)
            if XmeanN(counter) < 0
                XmeanN(counter) = 0;
                XstdN(counter)  = x(counter)./PHIinv(counter);
            end
        end
        % step 4
        dgdXnorm = eval(getGradient);

        for counter = 1 : length(dgdXnorm)
            if dgdXnorm(counter) == 0
                dgdXnorm(counter) = eps;
            end
        end
        % step 5
        dgdXnorm = dgdXnorm.*XstdN;
        state.gFNEval = state.gFNEval  + NbVariables + 1;
        gx = eval(getG);
        state.gFNEval = state.gFNEval  + 1;
        % step 6
x = (1/((dgdXnorm.')*dgdXnorm + eps*0))*((dgdXnorm.')*xstar -
gx).*dgdXnorm;
        % step 7
        aux = sqrt((x.')*x);
        DeltaBeta = beta - aux;
        beta = aux;
```

```
        stopCondition02 = DeltaBeta < BetaTol;

        % step 8
        x = XmeanN + x.*XstdN;
    end

    state.beta(iteration) = beta;

    gx = abs(eval(getG));
    state.gx(iteration) = gx;
    state.gFNEval = state.gFNEval  + 1;
    state.DeltaBeta = DeltaBeta;

    exitFlag = isItTimeToStop(options,state);

end

state.cputime = cputime;

Pf = (1 - normcdf(beta));                 % Eq. 3.8

state.distributions  = distributions; state.distParameters =
distParameters; state.XstdN = XstdN;

return
```

For a color version of this code, see www.iste.co.uk/radi/optimizations.zip

PART 2

Optimization

5

Combinatorial Optimization

5.1. Introduction

The term "combinatorial optimization" encompasses any optimization problem involving binary variables, together with appropriate techniques for solving such problems. The objective of combinatorial optimization is to choose a configuration that satisfies the conditions from the set of possible configurations for a given problem, while optimizing certain criteria. These criteria are generally represented as a function, known as the cost function or objective function. Currently, there is growing pressure to not only find a solution that satisfies the criteria and is reasonably efficient, but also to find one that is in fact the best possible solution. This is the case for decision problems in production management, logistics and telecommunications, where size increases rapidly and the financial stakes are huge.

In its most general form, a combinatorial optimization (or discrete optimization) problem finds the best feasible subsets (or solutions) from a discrete set. The concept of best solution is defined in terms of an objective function. Formally, given:

– a discrete set N;

– a set function $f\colon 2^N \mapsto \mathbb{R}$, called the objective function;

– and a set $\mathcal{R}$ of subsets of N whose elements are called feasible solutions;

the goal of a combinatorial optimization problem is to determine:

$$\max_{S \subseteq N}\{f(S)\colon S \in \mathcal{R}\}.$$

The set of feasible solutions cannot be listed exhaustively. Indeed, the difficulty of combinatorial optimization problems is that the sheer number of feasible solutions makes them impossible to enumerate, and even determining whether feasible solutions exist may be an NP-complete problem. The description of $\mathcal{R}$ is therefore stated implicitly. In general, this is the relatively short list of the properties of feasible solutions.

Most of the time, we are faced with one of the following special cases:

– $N = \{1, 2, \ldots, n\}$ $N = \{1, 2, \ldots, n\}$ and $\mathcal{R} = X \cap \mathbb{Z}^n$, where $X \subset \mathbb{R}^n$ and the cardinality of $\mathcal{R}$ is exponential with respect to n;

– $\mathcal{R} = conv(\mathcal{R}) \cap \mathbb{Z}^n$, so the properties of $\mathcal{R}$ translate to linear constraints, i.e. X is a polyhedron on $\mathbb{R}^n$;

– $f(S) = \sum_{s \in S} f(s)$.

A non-exhaustive list of combinatorial optimization problems might include the following:

– Loading problem: given a collection of objects, each characterized by a volume and a value, and a bag of limited volume, which objects should we load into the bag to maximize the total loading value?

– Traveling salesman problem (TSP): given n towns and the distances between them, which path should we take to visit every town exactly once while traveling the minimum possible distance?

– Covering problem: given a set of objects and a set of subsets of this set, each with a certain cost, which subsets should we choose to guarantee that every object is taken at least once while minimizing the total cost?

– Vehicle routing problem (VRP): given a set of clients that must each be visited exactly once to fulfill a certain demand, a fleet of delivery vehicles of limited capacity parked at the warehouse and the distances between each pair of clients and the distance between each client and the warehouse, which delivery routes should be followed in order to satisfy the demand of every client while minimizing the total distance traveled?

– Localization problem: there are several versions of this problem. One simple version can be stated as follows: if the goal is to minimize the total cost, how many warehouses should be opened and where should they be located given that (i) there are multiple different possible sites for opening a warehouse, each with a certain opening cost, and (ii) the warehouses must serve a set of clients, incurring costs that depend on which warehouse serves which client?

Combinatorial optimization is becoming increasingly important. From operations research to artificial intelligence, various solving methods have been proposed. These methods can essentially be classified into two families. The first family contains so-called exact, complete or optimal methods that guarantee admissible and optimal solutions, with respect to the function defined for the relevant problem. If such a solution does not exist, these methods must be able to conclude that there is no solution. The second family, known as approximate, incomplete or heuristic methods, has been proposed to mitigate combinatorial explosion at the expense of no longer guaranteeing optimality.

Complete methods use tree-based searches. They start from an empty solution and construct the optimal solution (if it exists) by ranging over the set of variables and instantiating them one by one. The approach taken by these methods can therefore be described as constructive. Backtracking is often necessary if the constraints are violated, which makes the search more complex. Despite much work on expanding the search tree, detecting inconsistencies at an earlier stage and/or reducing the amount of backtracking with heuristics, such methods remain exponential, meaning that they explode as the size of the problem increases.

Approximate methods, also known as heuristic methods, begin with a complete initial solution S_0, i.e. a solution where every variable has been instantiated. They then repair this solution step by step until a predefined stopping criterion is reached; the goal is of course to find a solution S^* which one hopes to be optimal with respect to the cost. These methods use an approach that can be described as local, neighboring or repair-based.

This chapter presents two very well-known problems of combinatorial optimization: the TSP and the VRP. The reader can refer to [TEG 12, GHE 07, KOR 11] for more details.

5.2. Symmetric TSP

Human beings face various problems across every part of society. A problem can be defined in terms of a set of properties that must be satisfied by its solutions. This might be a decision problem or an optimization problem.

In practice, in an optimization problem it is often desirable for some of its variables to be integers or even binary, $x_j = 0$ or 1. This is described as a discrete integer linear program (ILP), or a binary linear program in the binary case. One of the most widely studied problems in the class of ILP is the TSP, which was posed for the very first time by Euler (1736). Since then, it has been widely studied in the literature in its classical form. In 1859, it was stated as follows: a traveling salesman must visit finitely many towns exactly once, then return to his starting point. Find the order in which the towns should be visited in order to minimize the total distance traveled.

To better understand the problem, we can say that the traveling salesman should visit n given towns while passing through each town exactly once. He starts with an arbitrary town and ends by returning to the town in which he started. The distances between towns are known. Which path should he take to minimize the distance traveled? The notion of distance can also be replaced by other notions, such as the time or money that he must spend; in general, we can speak of a cost (see Figure 5.5).

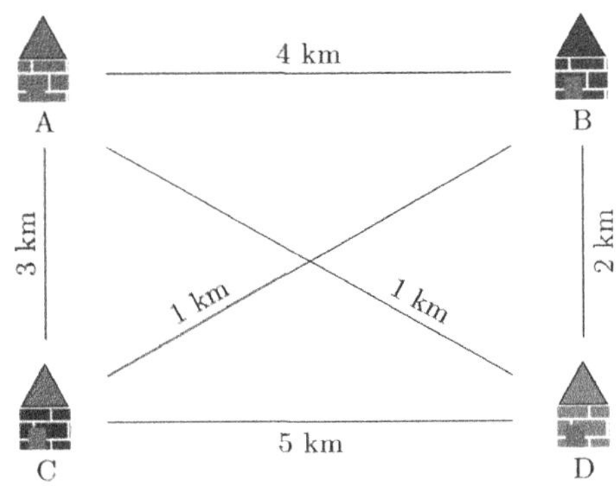

Figure 5.1. *Different possible trips. For a color version of this figure, see www.iste.co.uk/radi/optimizations.zip*

Which path is the shortest if the salesman starts at A, passes through every other point exactly once, then returns to A?

5.2.1. *Historical overview*

The origin of the name of the "traveling salesman problem" is not known exactly. It is one of the oldest combinatorial problems. William Rowan Hamilton made it into a game: Hamilton's Icosian game. Players had to choose a route passing through 20 points using only the predefined connections. In 1930, the TSP was studied in more depth by Karl Menger at Harvard, before being further developed at Princeton by the mathematicians Hassler Whitney and Merril Flood. Thus, a variety of mathematicians have taken up this problem since the early 20th century in search of a scientific answer to this very real question.

Tolga Bekta proposed several ways to solve the TSP [BEK 06]. In 1998, Arora [ARO 98] showed, based on a proof by Papadimitriou [PAP 77], that the Euclidean TSP belongs to the class of NP-hard optimization problems, which have a computational complexity that increases exponentially with respect to the number of towns in the problem. Dantzig *et al.* [DAN 54] solved a TSP problem with 49 towns using the cut generation method. Padberg and Rinaldi [PAD 91] solved a problem with 532 towns, followed by another problem with 2392 towns. In 2006, Applegate *et al.* [APP 06] solved a problem with 85,900 towns (see Figure 5.2). The renowned

Concorde program [APP 06] for solving the symmetric TSP has been the best-performing algorithm for the past 20 years. The program uses the branch-and-cut method, generating constraints from valid inequalities that have been specifically studied because of the TSP.

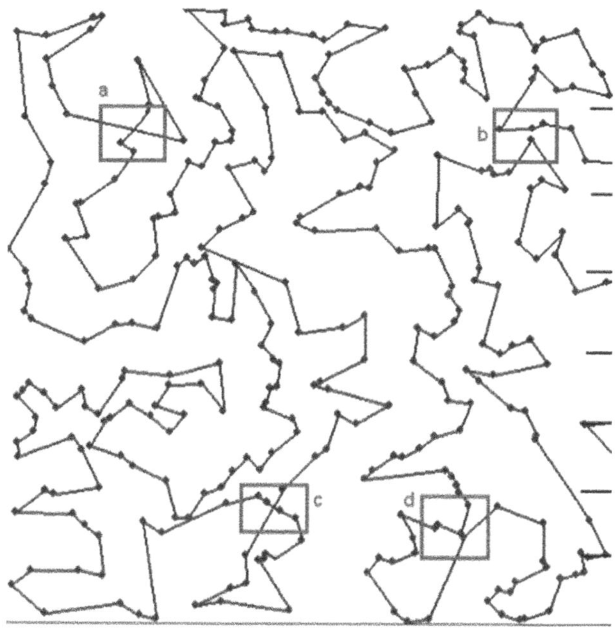

Figure 5.2. *Problem with 85,900 towns. For a color version of this figure, see www.iste.co.uk/radi/optimizations.zip*

The TSP can be divided into two types, each with several different variants: the symmetric traveling salesman problem (STSP) and the asymmetric traveling salesman problem (ATSP). This chapter will present the asymmetric case. In the ATSP, $d_{ij} \neq d_{ji}$, whereas in the STSP, $d_{ij} = d_{ji}$, where d_{ij} is the distance between two nodes i and j.

This problem has many practical applications and serves as a reference for other problems, particularly from the fields of transportation and logistics. Examples include finding the shortest path for a schoolbus to pick children up or the shortest distance traveled by the mechanical arm of a machine when punching holes into a circuitboard in an industrial setting. Many variants of the TSP can be obtained either by adding constraints – such as the probabilistic traveling salesman problem (PTSP) – or by modifying the problem, e.g. the VRP.

5.2.2. *Solving methods*

5.2.2.1. *Exact methods*

– Branch-and-bound: this method is based on "intelligently" enumerating the admissible solutions of a combinatorial optimization problem that involves finding a point that minimizes or maximizes a function, subject to several constraints. It is widely used to solve problems that are considered difficult to solve efficiently.

– Little algorithm: when the problem being solved involves binary variables, this algorithm can be used to simplify the computations. It is a simplification of the branch-and-bound algorithm (Algorithm 5.1).

Algorithm 5.1 Little algorithm

1: Reduce the matrix (row and column)
2: Select the empty cell that is most expensive to eliminate.
3: The elimination cost is the minimum extra cost incurred if the cell is not selected, i.e. if another cell on the same row and another cell on the same column are selected.
4: Remove the row of the starting town and the column of the destination town.
5: Set the return cost to infinity.
6: Restart with the subtable.
7: Present the algorithm as a tree (branch-and-bound).
8: Consider any path with lower cost than the first final cost that we find.
9: Follow the tree until it terminates.

5.2.2.2. *Approximate methods*

– Greedy algorithm: follows the principle of making a locally optimal choice step by step in the hope of obtaining a globally optimal result. For example, in the change problem (giving change with the fewest possible coins), the greedy algorithm simply chooses the biggest coin that does not exceed the remaining change to be given. In cases where the algorithm does not always find the optimal solution, it is called a greedy heuristic.

– Methods of descent: involves moving through the search space by always choosing the best solution among the neighbors of the current solution.

– Simulated annealing: an empirical (metaheuristic) method inspired by a process used in metallurgy that alternates between slow cooling cycles and heating cycles (annealing) to minimize the energy of the material. This method can be transposed to optimization to find the extrema of a function.

– Tabu search: an optimization metaheuristic. The idea is to explore the neighborhood around a given position and choose the position in this neighborhood that minimizes the objective function.

– Genetic algorithm: based on the principle of selecting individuals (routes) from a population (set of routes) with properties as close as possible to the desired properties. A selection criterion needs to be defined, as well as rules describing the evolution of the population. However, there are very few theoretical results involving genetic algorithms, and the choice of criteria and rules is therefore heuristic.

– Lin and Kernighan algorithm: one of the best heuristics for the TSP. The algorithm exchanges a certain number of edges from a given solution to find a solution with a better cost. This method exchanges two or three edges to make the route shorter.

– Ant colony algorithms: a family of metaheuristic optimization algorithms based on the behavior of ants and other species that form superorganisms. Proposed to find an optimal path in a graph, the first algorithm was inspired by the behavior of ants looking for a path between their colony and a food source.

Figure 5.3 summarizes the various solving methods cited above.

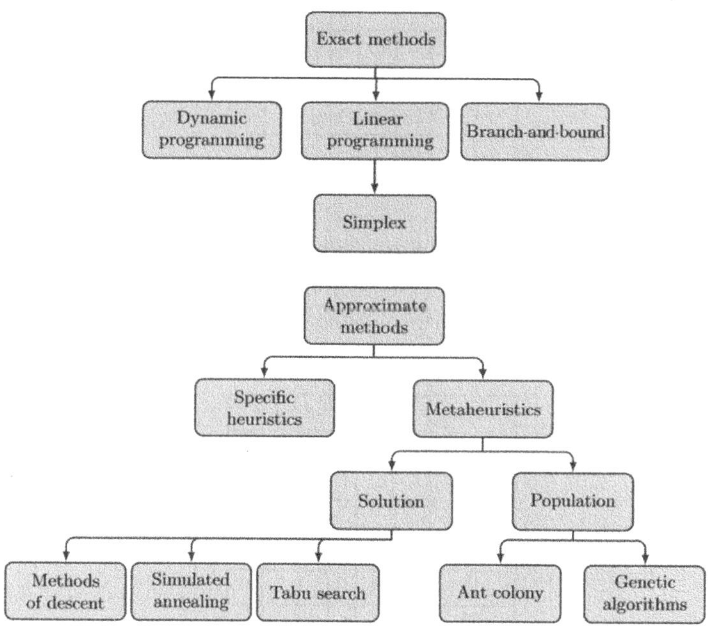

Figure 5.3. *Various solving methods*

Consider the following table listing transportation costs between towns:

Starting towns	A	B	C	D	E	F	Min
A	-	1	7	3	14	2	1
B	3	-	6	9	1	24	1
C	6	14	-	3	7	3	3
D	2	3	5	-	9	11	2
E	15	7	11	2	-	4	2
F	20	5	13	4	18	-	4

The first step of the method is to reduce the matrix by subtracting the smallest element from each row. This gives the reduced matrix illustrated in the following table:

Starting towns	A	B	C	D	E	F
A	-	0	6	2	13	1
B	2	-	5	8	0	23
C	3	11	-	0	4	0
D	0	1	3	-	7	9
E	13	5	9	0	-	2
F	16	1	9	0	14	-
Min	0	0	3	0	0	0

The table is easy to interpret: each transportation cost between town i and town j has been replaced by the difference between this transportation cost and the lowest transportation cost required to leave the town. The sum of the minimum costs is $1 + 1 + 3 + 2 + 2 + 4 = 13$.

Starting towns	A	B	C	D	E	F
A	-	0	3	2	13	1
B	2	-	2	8	0	23
C	3	11	-	0	4	0
D	0	1	0	-	7	9
E	13	5	6	0	-	2
F	16	1	6	0	14	-

This matrix can be interpreted as the extra cost of arriving in each town relative to the minimum cost of arriving, or alternatively as the table of transportation costs that need to be added to the minimum cost of $13 + 3 = 16$. To choose the first pair, we need to compute the penalties as follows: $a_{ij} = \min c_{ik} + \min c_{hj}$. The next table recalculates these penalties for each pair without the extra cost.

Starting	Destinations					
towns	A	B	C	D	E	F
A	-	0(2)	3	2	13	1
B	2	-	2	8	0(6)	23
C	3	11	-	0(0)	4	0
D	0(2)	1	0(2)	-	7	9
E	13	5	6	0(2)	-	2
F	16	1	6	0(1)	14	-

To choose an edge, we will take the pair B–E, because it has the largest penalty. Therefore, we need to delete row B and column E to go from a problem with six towns to a problem with five. This requires us to reduce the matrix and recalculate the penalties once again.

Starting	Destinations				
towns	A	B	C	D	F
A	-	0(2)	3	2	1
C	3	11	-	0(0)	0
D	0(2)	1	0(2)	-	9
E	13	5	6	0(2)	2
F	16	1	6	0(1)	-

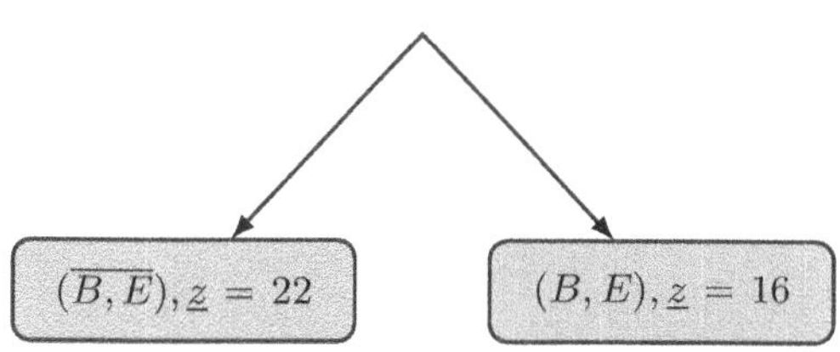

We will determine the pair (i, j) for which the new penalty is highest. There are two of them. As an example, we will branch at (D, A) as follows: $z = 16 + 3 = 19$.

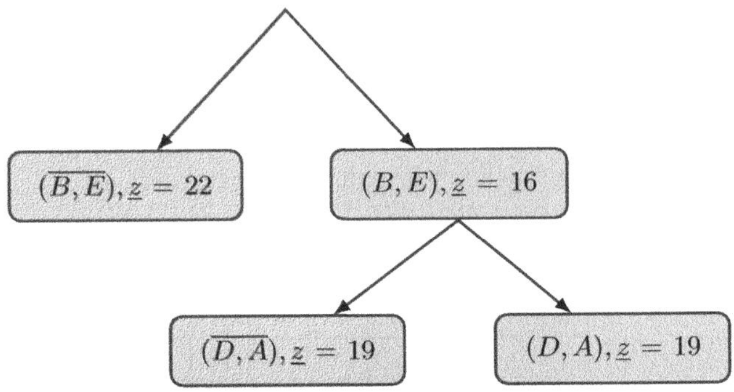

Repeating the same steps gives:

Starting	Destinations			
towns	B	C	D	F
A	0	3	-	1
C	11	-	0	0
E	-	6	0	2
F	1	6	0	-

After computing, we select the pair A, C. The minimum cost is therefore: $z = 19 + 3 = 22$.

Starting	Destinations			
towns	B	C	D	F
A	0(1)	0(3)	-	1
C	11	-	0(0)	0(1)
E	-	3	0(2)	2
F	1	3	0(1)	-

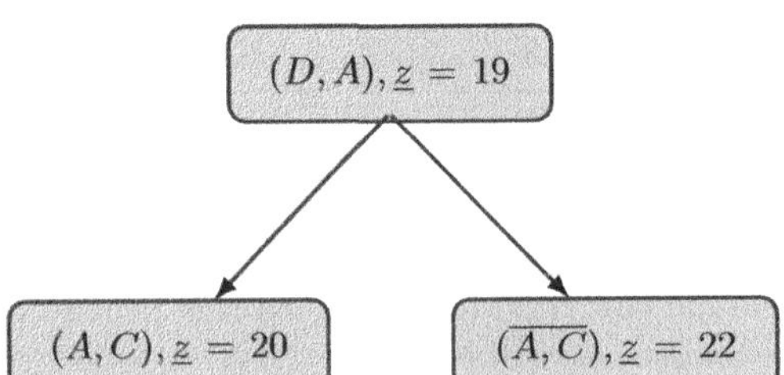

Next, we obtain the matrix that is shown in the following table:

Starting	Destinations		
towns	B	D	F
C	10	-	0
E	-	0	2
F	0	0	-

The minimum cost associated with this choice is therefore: $z = 19 + 1 = 20$.

Starting	Destinations		
towns	B	D	F
C	11	-	0
E	-	0	2
F	1	0	-

We reform the matrix by taking the last table into account:

Starting	Destinations				
towns	A	B	C	D	F
A	-	0	3	2	1
C	3	11	-	0	0
D	-	1	0	-	9
E	13	-	6	0	2
F	16	1	6	0	-

Now we need to reduce the matrix using the following table:

Starting	Destinations				
towns	A	B	C	D	F
A	-	0(2)	3	2	1
C	0(10)	11	-	0(0)	0(1)
D	-	1	0(4)	-	9
E	10	-	6	0(2)	2
F	13	1	6	0(1)	-

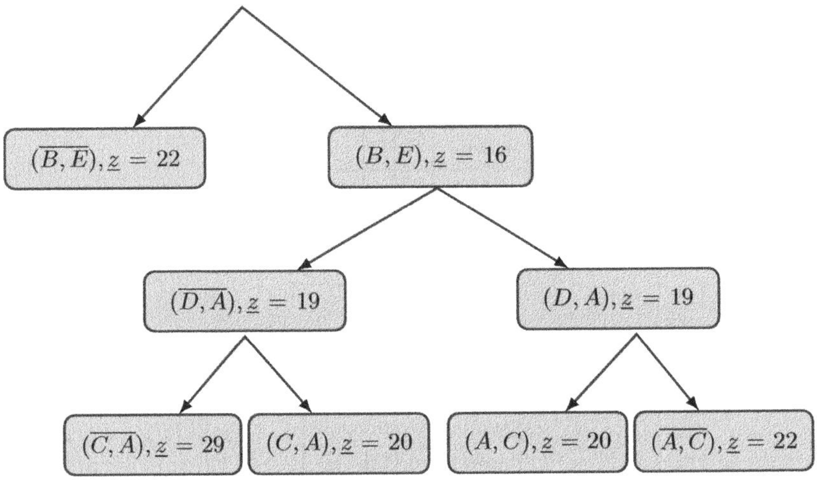

Finally, we need to eliminate the edge (A, C), then continue in the same way. We ultimately find that the optimal route is as follows: A, C, F, B, E, D with $z = 20$.

5.3. Asymmetric traveling salesman problem

The ATSP is defined on a directed graph $G = (V, E)$, where $V = \{1, \ldots, n\}$ represents the set of nodes (vertices), $\mathrm{E} = \{(\mathrm{i}, \mathrm{j})\colon \mathrm{i}, \mathrm{j} \in \mathrm{V}\}$ is the set of edges, and the non-symmetric cost matrix c_{ij} is defined on E. Solving the ATSP means finding a Hamiltonian cycle on G with the lowest cost. The problem is often interpreted as finding the optimal route of a traveling salesman through n towns. The ATSP is NP-hard even when the costs are Euclidean (see Figure 5.4).

5.3.1. *Variants of the ATSP*

Several variants of the ATSP can be formulated to describe it including:

– the periodic ATSP: in this generalization of the ATSP, we define a schedule over a given period of k days. Thus, we want to find k cycles in G such that each node of G is visited a certain number of times over the k-day period and the total cost of the route over the k-day period is minimal [PAL 04];

– ATSP with replenishment arcs (RATSP): the goal is to find a route of minimal cost. In this context, we allow two parallel edges to be defined between the same two nodes, namely a replenishment edge and an ordinary edge, provided that the ordinary edge has lower cost. The RATSP is a generalization of the ATSP with various applications in routing problems. For example, it is used in the aircraft routing problem, where maintenance options are modeled as replenishment arcs, as well as the

VRP, where customer demand and vehicle capacity can be modeled as replenishment arcs [BOL 00];

– precedence-constrained ATSP: also known as the sequential ordering problem (SOP), the precedence-constrained ATSP requires certain nodes to be preceded by others in any feasible route. This problem is an essential model in scheduling and routing and has a wide range of applications, including vehicle routing, sequencing in manufacturing industries, and so on [SAR 05, GOU 06];

– ATSP with time windows (ATSP-TW): this is a classical scheduling and demand pathing problem. The idea is to impose time constraints specifying when each vertex may be visited [ASC 01];

– ATSP with jump constraints (Hamiltonian TSP–HTSP): let S be the set of towns subject to the jump constraint (i.e. whose distance in terms of the number of edges between towns is at least s). The HTSP searches for a Hamiltonian cycle of minimal cost that observes the distance constraints between the towns of S;

– probabilistic ATSP with clients: here, we assume that each client is visited with probability p_i, independently of the others (variant of the classical ATSP).

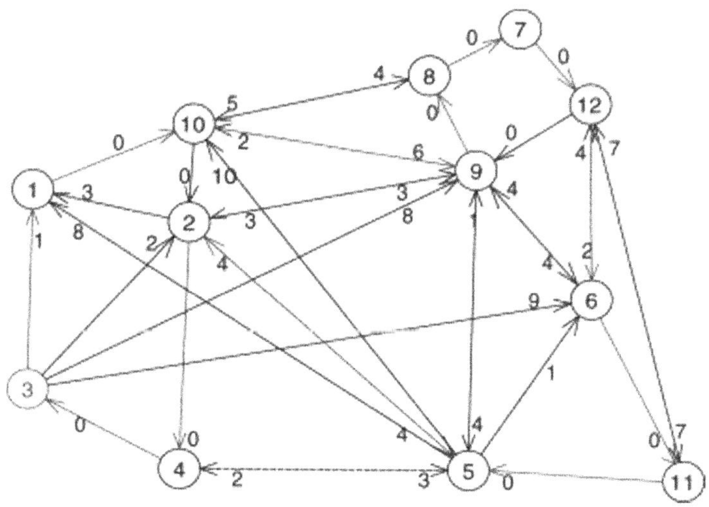

Figure 5.4. *Traveling salesman's route through n towns*

In general, the ATSP is known to be an NP-hard problem. It has been studied intensively over the past six decades. Early studies on the topic were presented by Fischetti *et al.* [FIS 94] and Oncan *et al.* [ÖNC 09]. Several books discussing the TSP and its variants have been published. Examples include the books by

Lawler *et al.* [LAW 85], Reinelt [JÜN 95], Gutin and Punnen [GUT 06] and Applegate *et al.* [APP 06].

5.3.2. *Mathematical formulations*

Let $G = (V, A)$ be a complete and directed graph, where the set V represents the towns to be visited, including the starting town of the route, and A is the set of edges of G. Each edge in A is assigned a travel distance $d_{\vec{p}}$ from town i to town j. The length of a path in G is the sum of the distances of the edges on the path. Thus, the TSP reduces to find a Hamiltonian cycle (closed path passing through every vertex of the graph exactly once) of minimal length in G. In general, the TSP can equivalently be formulated by assigning each pair of towns to be visited ($i = 1$ to n, $j = 1$ to n and $i \neq j$), a binary succession variable that takes the value 1 if town j is visited immediately after town i in the route and 0 otherwise, with the set of successors of i. The basic idea of the mathematical formulation of the TSP is to begin with an assignment problem (AP): each town i is assigned a town j that should succeed it on the salesman's route. This gives the formulation:

$$\min \sum_{i=j} d_j x_j \qquad [5.1]$$

$$\text{s.t.} \sum_{j \in N^+(i)} x_{ij} = 1 \quad \forall i \in V \qquad [5.2]$$

$$\sum_{i \in N^+(j)} x_{ij} = 1 \quad \forall j \in V \qquad [5.3]$$

$$x_i \in \{0; 1\}, \forall (i, j) \in A \qquad [5.4]$$

The first two constraints, [5.2] and [5.3], translate the fact that each town must be visited exactly once, and [5.4] is an integrality constraint. This formulation is not yet valid in the sense that a solution does not necessarily give a full cycle: in general, it consists of several cycles (subtours). We therefore need to extend this formulation in such a way that cycles (subtours) are prohibited. There are two ways to do this classically, proposed by Dantzig *et al.* [DAN 54]. Miller *et al.* [MIL 60] rewrote the problem differently to obtain two formulations of the TSP: the DFJ (Dantzig, Fulkerson and Johnson) formulation and the MTZ (Miller, Tucker and Zemlin) formulation. A few years later, Desrochers and Laporte reinforced the MTZ formulation by proposing the DL formulation [DES 91].

5.3.2.1. *DFJ formulation*

Here, A represents the set of edges (i, j) and V is the set of towns, with n being the number of towns to be visited. The value l_{ij} represents the distance between i and j. We also use a binary variable: for each edge $(i, j) \in A$, the variable x_{ij} is 1 if the edge (i, j) is taken and 0 otherwise, with S the subset of vertices that have been visited (in subtours) [DAN 54]:

$$\min \sum_{i \neq j} d_{ij} x_{ij} \qquad [5.5]$$

$$\text{s.t.} \sum_{j=1}^{n} x_{ij} = 1 \quad (\forall i \in V, i \neq j) \qquad [5.6]$$

$$\sum_{i=1}^{n} x_{ij} = 1 \quad (\forall j \in V, j \neq i) \qquad [5.7]$$

$$\sum_{i,j \in S} x_{ij} \leq |S| - 1 (S \subset V, 2 \leq |S| \leq n - 2) \qquad [5.8]$$

$$x_{ij} \in \{0; 1\}, \ \forall (i, j) \in A \qquad [5.9]$$

The constraint [5.5] represents the objective function, [5.6] and [5.7] represent the assignment constraints, [5.8] represents the subtour elimination constraints and [5.9] is the integrality constraint.

This formulation proves to be difficult to manipulate in practice due to the size of the model as the number of towns increases. This increase is linked, in particular, to the exponential number of subtour elimination constraints that are generated. This formulation of the TSP contains $n(n-1)$ variables and $\left(2^{n-1} + n - 2\right)$ constraints.

Although it is easy to express, the primary shortcoming of this method is its size, which quickly becomes prohibitively large. Indeed, for an n-dimensional problem, there is a very large number of subtour elimination constraints, which quickly renders problems unsolvable as the number of clients to be visited increases. Alternative methods that have been developed therefore tend to reduce this number of constraints by relaxing the problem. This, for example, was done in 1960 by Miller *et al.* [MIL 60] by adding an additional variable to reduce the final number of constraints. However, this method yields results that are less optimal than the previous method, although obtained more quickly and for problems with larger dimensions. Rather than using an integer linear program to solve these formulations, algorithms were developed to find or approximate the optimal solution more quickly and with fewer resources. The most widely used algorithm is the branch-and-bound algorithm, which separates the main problem into subproblems and evaluates the solutions of each subproblem to reconstruct the solution of the initial problem.

5.3.2.2. *MTZ formulation*

A represents the set of edges (i, j) and V the set of towns, with n being the number of towns to be visited. The value l_{ij} represents the distance between i and j. We also use a binary variable x_{ij}: for each edge $(i, j) \in A$, the variable x_{ij} is 1 if edge (i, j)

is taken and 0 otherwise. The quantity U is the visiting rank of i or j and d is the distance traveled [MIL 60]:

$$\min \sum_{i \neq j} d_{ij} x_{ij} \qquad [5.10]$$

$$\sum_{j=1}^{n} x_{ij} = 1 \quad (\forall i \in V, i \neq j) \qquad [5.11]$$

$$U_i - U_j + n x_{ij} \leq n - 1, \qquad [5.12]$$

$$U_j = U_i + 1 \ \text{ if } x_{ij} = 1, i \neq j \qquad [5.13]$$

$$x_{ij} \in \{0; 1\}, \ \forall (i, j) \in A \qquad [5.14]$$

The constraint [5.10] is the objective function, [5.11] and [5.12] are the assignment constraints, [5.13] is the integrality constraint and [5.14] represents the MTZ subtour elimination constraints. This formulation of the TSP has $(n - 1) \times (n + 1)$ variables and $\left(n^2 - n + 2\right)$ constraints.

5.3.2.3. *Desrochers–Laporte formulation*

This is a reinforced version of the MTZ formulation [DES 91]:

$$\min \sum_{i \neq j} d_{ij} x_{ij} \qquad [5.15]$$

$$\text{s.c.} \sum_{j=1}^{n} x_{ij} = 1 \quad (\forall i \in V, i \neq j) \qquad [5.16]$$

$$\sum_{i=1}^{n} x_{ij} = 1 \quad (\forall j \in V, \mathrm{j} \neq i) \qquad [5.17]$$

$$U_i - U_j + (n - 1) x_{ij} + (n - 3) x_{jt} \leq n - 2_{2n} \ \forall i \neq 1, j \neq 1, \qquad [5.18]$$

$$2 \leq U_i \leq n, \ U_1 = 1 \qquad [5.19]$$

$$x_{ij} \in \{0; 1\}, \ \forall (i, j) \in A \qquad [5.20]$$

The constraint [5.15] represents the objective function, [5.16] and [5.17] are assignment constraints, each town is visited exactly once, [5.18] is an integrality constraint and [5.19] is the DL subtour elimination constraint.

5.3.3. ***Methods for solving the ATSP***

Methods for solving the ATSP are generally split into two categories: exact methods and approximate methods (heuristics and metaheuristics). Designed to find the optimum of a problem, exact methods often have computation times that increase exponentially with the size of the problems that they are attempting to solve; this means they must be terminated prematurely during resolution. If so, the exact method only produces a lower or upper bound. By contrast, approximate methods are designed to produce high-quality admissible solutions within a reasonable resolution time, although these solutions are not necessarily optimal. Approximate methods encompass constructive heuristics, local searches that give a sequence of solutions with decreasing (or increasing) costs until the local optimum is attained, and finally metaheuristics that use various mechanisms to escape from local optima. Exact methods and approximate methods are complementary as their results allow them to be evaluated against one another: a small gap between lower and upper bounds means that the optimum is nearby, whereas a gap of zero proves that the values found are optimal. As a result of this complementary relationship between exact methods and approximate methods, one method is often used to improve the performance of another [FIS 94].

Solving an ILP requires the application of dedicated methods such as branch-and-bound methods [LAW 85, NAR 77], branch-and-cut methods [PAD 91, PER 15], column generation methods, [DES 06, VAN 96], dynamic programming [BEL 56, BER 95] and Lagrangian relaxation [MAC 83].

Exact algorithms are used to find at least one optimal solution to a problem. The most successful exact algorithms in the literature follow the paradigms of four large classes of solving methods:

– dynamic programming;

– constraint programming;

– implicit enumeration methods (branch-and-bound);

– cutting-plane methods.

These methods use bounds to eliminate sets of solutions with a certain structure from the search. Other methods are specific to certain problems or less general, such as dynamic programming and the cutting-plane method.

5.3.3.1. *Branch-and-bound method*

ATSPs are often solved optimally with the branch-and-bound method, which considers a relaxed problem called the AP [FIS 94]. Instead of having a unique route through every town (node), an optimal solution of the AP generally consists of more than one route, called a subtour. The branch-and-bound algorithm is constructed from the four following steps.

The branching rule specifies how the current problem should be subdivided into subproblems. The idea is to perform a separation, which requires us to choose a separation variable. This variable is chosen heuristically, with various possible choices that may depend on the effectiveness of the solving method. The pathing strategy through the tree determines the next subproblem that will be considered. There are two widely used strategies, depth-first search (DFS), which starts by solving the last subproblem to be generated, and breadth-first search (BFS), which first solves the subproblem judged to be best, typically the subproblem with the lowest AP bound. The upper bound strategy determines how the routes should be constructed in the branch-and-bound process. One possible method for constructing a feasible route for the ATSP, based on AP, solutions is the readjustment procedure.

The lower bound strategy determines how the lower bound associated with the value of a solution should be constructed for a given subproblem. The value of the AP solution of the subproblem is typically taken as a lower bound. A review of current knowledge about the branch-and-bound algorithm for the ATSP problem can be found in Miller and Pekny (1991). The algorithms presented by Miller and Carpaneto apply a readjustment to obtain upper bounds using the AP lower bounds, branching on the smallest cycle in the current AP solution. The pathing strategy used by the two algorithms is BFS (best subproblem first), which means solutions are obtained very quickly for many real examples of the ATSP. It also means that a list of subproblems must be maintained to determine the most promising subproblem. Consequently, BFS algorithms tend to encounter memory issues as the pathing tree grows. The upper bounds of the branch-and-bound algorithms for the ATSP were improved by Turkensteen in 2007 with an iterative readjustment, a constructive procedure that yields good feasible solutions for the ATSP. The iterative readjustment procedure allows the number of subproblems in the branch-and-bound tree to be reduced in such a way that the overall computational time is smaller [LAW 85].

Algorithm 5.2 Branch-and-bound algorithm

Start
Add the starting node of length 0 to a list.
Repeat
If the first branch contains the desired node then
End successfully.
Else
Delete the branch from the list and form new branches by extending the deleted branch by one step.
Compute the cumulative costs of the branches and add them to the list, sorted by increasing order.
Until (list empty or desired node is found)
End

5.3.3.2. *Cutting-plane method*

The cutting-plane method is an alternative to the branch-and-bound algorithm that is also used to solve integer-valued problems. The fundamental idea is to add suitable constraints called cuts to a linear problem until the optimal realizable basic solution becomes integer valued. Each cut, with respect to a current partial solution, satisfies the following criteria:

1) every integer-valued feasible solution is feasible with the cut;

2) the current partial solution is not feasible with the cut.

There are various ways to generate cuts. Examples include Gomory cuts, which produces cuts for any table of a linear program. This has the advantage of being applicable to any problem, but the disadvantage of being very slow. The second approach is to use the structure of the problem to produce very high-quality cuts. This requires a case-by-case analysis but can lead to very effective solving techniques.

5.3.3.3. *Branch-and-cut method*

Padberg and Rinaldi [PAD 91] improved the idea of the branch-and-bound algorithm for linear programming (LP) by describing a method that uses inequalities to reinforce LP relaxation (inequalities valid on the convex envelope of integer solutions). They called their method the branch-and-cut method. Since then, many implementations have used this approach.

This method requires the optimization problem to be transformed into an integer LP problem. Unfortunately, it is difficult or even impossible in general to enumerate every inequality that follows from the optimization problem, even if we restrict attention to the inequalities that describe the convex envelope around the optimal region. Once the problem has been transformed, if this is possible, we take advantage of the fact that the real-valued LP problem is easier to solve than the integer-valued problem. The branch-and-cut algorithm uses this specific property to accelerate the search process. The basic principle is the same as the branch-and-bound algorithm, but instead of "branching" on a variable to descend in the search tree, we can instead look for "cutting planes" to restrict the space of feasible solutions. At the start of the algorithm, we restrict attention to a small set of inequalities of the problem that are solved with real numbers. Then, if the optimal solution obtained for the real-valued problem is:

– an integer: we cut the branch being considered, updating the value of the optimal solution if necessary;

not an integer: we apply a separation algorithm and/or heuristics to extract a set of inequalities that are violated. If this set is empty, we search for an integer solution with this cost, otherwise we update the set of inequalities and continue the algorithm. The goal is to "intelligently" exclude the non-integer optimal solution from the search space using cutting planes.

If the problem being solved does not have a feasible solution, we cut the branch.

There are various ways to generate cutting planes, such as by choosing a linear combination of the inequalities of the LP [GOM 63], using Dantzig's simplex algorithm, or in a manner that is specific to the problem being solved. If we fail to generate a cutting plane and the subproblem cannot be pruned due to the value of the objective, we need to "branch". The branching operation is performed by specifying a set of hyperplanes that divide the current subproblem in such a way that the current optimal real solution is not feasible in any of the LP relaxation subproblems that are generated (for example, if the optimal real solution x_i is 2.7, we can branch by adding the constraint that $x \leq 2$ in one branch and $x \geq 3$ in the other).

Algorithm 5.3 Branch-and-cut algorithm

/* We want to solve the combinatorial optimization problem $\min\{c^t x : Ax \geq b, x \in \mathbb{R}^n\}$ With $A \in \mathbb{R}^{m \times n}$ and $b \in \mathbb{R}^m$.
List-of-problems = empty;
Initialize the linear program with the subsystem of constraints (A1; b1)
with $A_1 \in \mathbb{R}^{m_1 \times n}$ and $b_1 \in \mathbb{R}_1^m$ with $m_1 \ll m$
/* Problem evaluation steps */
Calculate the optimal solution x of the linear program. Current-solution = Apply-the-method-of-polyhedral-cuts ();
/* End evaluation steps */
If Current solution is feasible **then**
$x^* = \bar{x}$ is the optimal solution of $\min\{c^t x : Ax \geq b, x \in \mathbb{R}^n\}$
Else
Add the problem to List-of-problems;
End If
While List-of-problems is not empty **do**
Select a problem;
Branch the problem into two or more subproblems;
Solve the new subproblems (by applying the evaluation steps);
End While

5.4. Vehicle routing problem

5.4.1. *Definition*

The general problem of constructing routes for vehicles is known as the VRP. This is a multiobjective combinatorial optimization problem that has been widely studied and presented in many different variants in the literature. The VRP was introduced by Dantzig and Ramser, who called it the truck dispatching problem [DAN 59]. It belongs to the category of NP-hard problems [CHR 79, LEN 81, SAV 95].

The principle of the VRP is as follows: given a warehouse D and a set of customer orders $C = \{c_1, \ldots, c_n\}$, construct a set of routes for finitely many vehicles that start and end at a warehouse. In these routes, each customer must be served by exactly one vehicle, and the transportation capacity of a vehicle must not be exceeded in a route [NAB 07]. The objective of the VRP is to minimize the sum of the distance traveled or the overall travel time of the vehicles' routes, while satisfying the customer demand.

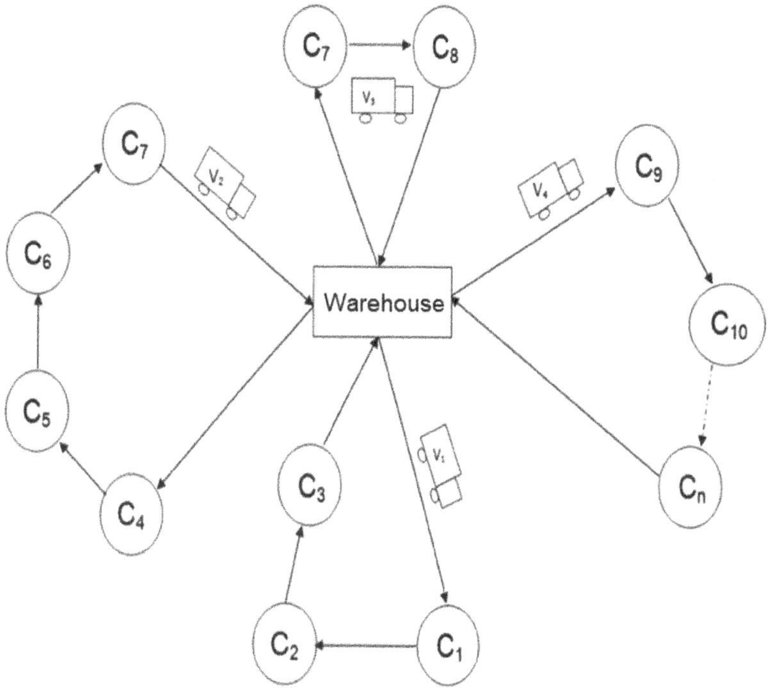

Figure 5.5. *VRP*

5.4.2. *Fields of application*

The VRP has many diverse applications. Every industrial company wants to improve the efficiency of its logistical chain to reduce the cost of producing goods or services, and to deliver its products smoothly. The VRP is a key link in the field of logistics. Given the massive increase in the cost of energy, almost 30% of the cost of producing a product or a service is due to transportation.

The VRP is part of our everyday lives, starting with picking schoolchildren up, picking up employees, collecting household waste, harvesting raw milk, distributing newspapers, mail and food products such as milk, bread and water. Ambulance services and urgent distribution of pharmaceutical products are also examples of the VRP.

In the last two examples, customers are served according to priority and urgency, and the objective is not to minimize the total distance traveled but to serve as many clients as possible in a given time frame. Note that all of these problems are constrained by time, and collection problems have more rigid time constraints than the others, which are distribution problems.

Information technology and robotics are other fields in which the VRP is widely used. Indeed, routing information in remotely connected computer networks or over wires is fundamentally based on the VRP. Aircraft scheduling is another example of an application.

Despite the interest that people have shown in this problem, to this day we have been incapable of solving some instances of it in practice. Large-scale instances of the VRP have proven to be difficult or complex and sometimes even "impossible" to solve with a calculator, regardless of its computational power.

5.4.3. *Parameters of the VRP*

The VRP is characterized by the network of the problem, the clientele that needs to be served, the fleet of vehicles and the objective function:

– *Network*: the transportation network allows the flow of individuals, freight or information. In a certain sense, it provides the skeleton of a system that aims to establish some form of communication. It can be represented as a complete asymmetric or symmetric graph. The vertices represent the clients, characterized by their geographical positions (the coordinates (x, y)) and the edges represent the paths between clients. With some constraints, this graph can be directed.

– *Clientele*: each client is characterized by a demand for a service or products (goods) of one or several types. The total demand of the clients on a single route must not exceed the capacity Q of the vehicle. Clients are also characterized by a position in space. Finally, the demand can be deterministic (fixed quantity demanded by the client that is known by the distributor) or uncertain (stochastic).

– *Fleet of vehicles*: the first criterion of the fleet is its size (number of vehicles), the second is its homogeneity (the vehicles are characterized by the same capacity and transportation cost) or heterogeneity (the vehicles have different capacities and/or transportation costs).

– *Objective function*: the most common objectives are to minimize the number of vehicles used or minimize the total distance traveled by the vehicles.

Other objectives can also be considered:

– minimizing the total time taken to complete the route, the waiting time, the delay, the service time, etc.;

– minimizing the number of vehicles;

– minimizing the total cost of the route. In the fixed cost case, we have amortization of equipment (vehicles and otherwise), driver salaries, vehicle fees, etc. In variable cost problems, late penalties (especially in the VRP-TW), etc.;

– maximizing the service quality;

– maximizing loading of the vehicles on routes.

5.4.4. *Variants of the VRP*

Several variants of the VRP can be defined by varying its parameters or by deleting, adding and/or combining the constraints of the classical problem. A classification of the variants of the VRP according to type of constraint is as follows:

– VRPC (vehicle routing problem with capacities): this is a routing problem with capacity constraints. The vehicles have a limited capacity (quantity, volume, weight, etc.). This is translated by the constraint that the sum of client demands on a given route must not exceed the vehicle capacity. In practice, every VRP is a limited capacity problem; only the extent of the influence of this constraint varies. For example, problems involving the transportation of goods are very sensitive to capacity, whereas the mail distribution problem is less sensitive.

– VRPCC (vehicle routing problem with complete capacity): this is a routing problem that requires the vehicle's capacity to be completely utilized.

– VRPHF (vehicle routing problem with heterogeneous fleet): the fleet is made up of vehicles of different types that are distinguished by their capacity, power, transportation cost, etc.

– OVRP (open vehicle routing problem): identical to the VRP except the vehicle has a free choice of whether to return to the warehouse after completing its route. If the vehicle chooses to return to the warehouse, it must travel the same route in reverse.

– RVRP (return vehicle routing problem): in this problem, the vehicle is required to return to the warehouse. The vehicle must return to the warehouse immediately after serving the last client, without needing to repeat its route.

– VRP with deterministic demand: frequently encountered by companies that deliver on request. The delivery driver knows the quantity of goods that must be delivered to each client before leaving the warehouse.

– VRP with stochastic demand: unlike the previous problem, the delivery driver does not know the quantity of goods to deliver to each client. This quantity is only revealed at the time of delivery. The demand of each client is estimated approximately with a stochastic function.

– SVRP (split delivery vehicle routing problem): involves visiting clients several times to fully satisfy demand. An exception for this problem, the client's demand can exceed the vehicle capacity.

– VRPCD (vehicle routing problem with collection and delivery): this problem extends the VRP to both collection and delivery. With this type of problem, the service time is counted twice, since both a collection and a delivery must be performed or vice versa.

– VRPMW (vehicle routing problem with multiple warehouses): vehicles can be supplied by multiple warehouses.

– VRP1W: vehicles are supplied by a single warehouse.

– VRPMP (vehicle routing problem with multiple products): a range of products must be delivered to different clients by each vehicle in a single route.

– VRP (vehicle routing problem with a single product): a single product must be delivered to different clients by each vehicle in a single route.

– PVRP (periodic vehicle routing problem): this problem has it that each client is periodically visited according to a predefined schedule.

– VRP with deterministic service time: the service time is known by the delivery driver before starting the route.

– VRP with stochastic service time: the delivery driver does not know the service time, which is only revealed when clients are served. A stochastic function is defined to approximate the service time.

– VRP-TW (vehicle routing problem with time windows): this extends the VRP to include time windows. Each client must be served in a certain time window that is known to the delivery driver in advance, and any violation of this constraint may incur a penalty. If the time window constraint is not satisfied, either the solution is rejected in the rigid version of the problem, or a penalty function is constructed and added or otherwise combined with the objective function in the relaxed case. In practice, this is a very common problem. The distribution of perishable products (milk, meat, etc.), newspapers, ambulance services, etc., are practical examples of the VRP-TW. In this class of problems, we can distinguish between two subclasses: the rigid VRP-TW, where the service must necessarily be provided within the time window, and the relaxed VRP-TW, where only being early or late incurs a penalty. Adding this penalty defines a variant of the VRP-TW.

– Static VRP: a problem where every component is known and fixed before beginning any route. This case is obtained either by a rigorous stationarization study of the set of parameters or because the problem being modeled describes a static phenomenon.

– Dynamic VRP: as opposed to the static VRP, the dynamic VRP has at least one dynamic or variable component at the time of execution. For example, the demand or the service time window might vary, or the number of clients to be served might change. This makes the problem more complex.

– SVRP (stochastic vehicle routing problem): this represents a special case of the dynamic VRP where the variable component follows a probabilistic law whose variation is described by a mathematical model, allowing this variation to be manipulated, making the problem easier to solve.

5.4.5. *Mathematical formulation of the VRP*

There are various formulations of the VRP in the literature. An aspect shared by all of them is that the VRP is represented as a directed graph. The graphical representation of the classical VRP is as follows: let $G = (X, U)$ be a directed graph, where:

– X is the set of vertices of the graph, G is the set of clients and warehouse(s) of the VRP, $X = N \bigcup \{0\}$, and $(|X| = n+1)$; U is the set of edges of the graph, G represents the paths between clients and between clients and warehouses, $(|U| = \frac{(n+1)n)}{2})$;

– weights can be assigned to the vertices (or edges) to define the quantities demanded by clients (or the distance separating two clients, or the displacement time, etc.).

There are many mathematical formulations of the classical VRP in the literature. The two formulations presented below are among the most commonly used. First, we must define the following parameters:

– n: number of clients; m: number of vehicles $(m \leq n)$;

– Q_k: capacity of the kth vehicle; T_k: maximum time of the route of the kth vehicle;

– as_j: number of edges leaving the vertex j; ae_j: number of edges entering the vertex j;

– q_j: demand of the jth client $d_0 = 0$;

– sk_j: service time of the jth client by the kth vehicle $(sk_0 = 0)$;

– t_{ij}: time needed to go from the ith client to the jth client $(t^k_{jj} = +\infty)$;

– d_{ij}: distance between the ith client from the jth client ($d^k_{jj} = +\infty$);

– C_{ij}: cost of moving from the ith client to the jth client.

Formulation

The most widely used formulation of the classical VRP in the literature is the formulation adopted by Laporte [LAP 92], Toth and Vigo [TOT 01], and Crainic and Semet [CRA 14]. It requires us to define $n \times n$ binary decision variables with the three following indices:

$$x^k_{ij} = \begin{cases} 1 \text{ if the edge (i,j) is traveled by the } k\text{-th vehicle,} \\ 0 \text{ otherwise,} \end{cases} \qquad [5.21]$$

In other words: $x^k_{ij} \in \{0, 1\}$; $i = 0, \dots, n,\ j = 0, \dots, n,\ k = 0, \dots, m.$

The problem is modeled as follows:

$$\min \sum_{i=1}^{n} \sum_{j=1}^{n} c_{ij} \sum_{k=1}^{m} x^k_{ij} \qquad [5.22]$$

$$\sum_{i=1}^{n} \sum_{k=1}^{m} x^k_{ij},\ j = 1, \dots, n \qquad [5.23]$$

$$\sum_{j=1}^{n} \sum_{k=1}^{m} x^k_{ij},\ i = 1, \dots, n \qquad [5.24]$$

$$\sum_{i=1}^{n} x^k_{ijp} - \sum_{k=1}^{m} x^k_{ipj} = 0; k = 1, \dots, m,\ p = 1, \dots, m \qquad [5.25]$$

$$\sum_{i=1}^{n} q_i \left(\sum_{j=1}^{n} x^k_{ij} \right) \leq Q_k,\ k = 1, \dots, m \qquad [5.26]$$

$$\sum_{j=1}^{n} s^k_i x^k_{ij} + \sum_{i=1}^{n} \sum_{j=1}^{n} t^k_{ij} x^k_{ij} \leq T_k,\ k = 1, \dots, m \qquad [5.27]$$

$$\sum_{j=1}^{n} x^k_{0j} \leq 1,\ k = 1, \dots, m \qquad [5.28]$$

$$\sum_{i=1}^{n} x^k_{i0} \leq 1,\ k = 1, \dots, m \qquad [5.29]$$

$$\sum_{i \in S} \sum_{j \in S} x^k_{ij} \leq |S| - 1,\ \forall S \subset N \text{ et } 2 \leq |S| \leq (n-1). \qquad [5.30]$$

Formula [5.22] represents the function to be optimized or the objective function of the classical VRP. In general, the goal is to find the minimum of the global routing cost. Formulas [5.23]–[5.30] represent the constraints of the problem.

The constraints [5.23] and [5.24] guarantee that each client is only served once and only by a single vehicle. The constraint [5.25] guarantees the continuity of the route: not only must each vehicle pass exactly once through a client, [5.24] and [5.25], the vehicle must necessarily also leave this client once the service is complete.

The constraint [5.27] guarantees that the capacity of the vehicle is not exceeded. The constraint [5.28] guarantees that the total duration of a route never exceeds its total maximum duration.

The constraints [5.28] and [5.29] guarantee that vehicle availability is not violated by ensuring that each vehicle only leaves and returns to the warehouse once. The final constraint [5.30] eliminates any subtours.

5.4.6. *Algorithmic complexity*

Evaluating the complexity of an arbitrary problem means estimating the number of instructions that must be performed in the worst-case scenario, to solve an instance of the problem. The algorithmic complexity measures the algorithm's efficiency at solving the problem. The higher the number of instructions, the more complex the problem. Thus, the algorithmic complexity is proportional to the number of instructions in the problem. Similarly, the more complex the algorithm, the less efficient it is.

We can distinguish between several types of complexity, such as constant complexity $O(1)$, which is independent of the problem size, logarithmic complexity $O(\ln(n))$, linear complexity $O(n)$, quadratic complexity $O(n^2)$, cubic complexity $O(n^3)$, polynomial complexity $O(n^p)$, exponential complexity, and finally factorial complexity $O(n!)$.

The complexity criterion allows us to distinguish between four classes of problems as follows:

– P class: this class contains all problems that can be solved in polynomial time by a Turing machine. Solving any instance of a problem in this class can be done in polynomial time. This class is known as the class of polynomial problems or P class.

– P-complete class: this class contains all polynomial problems and non-polynomial problems that can be reduced in poly-logarithmic time with or without the use of a deterministic calculating machine.

– NP class: this class contains all problems that can be solved in polynomial time on a Turing machine.

– NP-hard class: this class contains all problems whose complexity is exponential or doubly exponential. It is characterized by the difficulty of solving large problem instances exactly or optimally in polynomial time. Instead, we must be satisfied with so-called high-quality solutions in polynomial time. The problems of this class include tree partitioning problems, connectedness problems in a routing graph, etc.

The VRP is a routing problem. Lenstra and Rinnooy Kan [LEN 81] showed that the VRP is NP-hard. Consequently, it is not appropriate to attempt to solve large problem instances with exact methods. Instead, we must solve these with heuristic approaches that yield feasible and useful solutions in a reasonable time frame.

5.5. Selective routing problem

Faced with the current increase in energy prices, the transportation sector is showing a growing interest in this problem. Indeed, to minimize transportation costs, companies regularly seek to streamline their logistics systems to remain competitive. Logistics systems consist of various elements, one of the most important of which is the vehicle routing schedule in their transportation networks.

Over the past century, several models and methods have been proposed to answer the various problems encountered in logistics systems. These techniques are directly linked to the evolution of computer systems. Today, in just a few seconds, with a calculator of average capacity, we can solve logistics problems that would have been completely impossible to solve historically.

For example, we can study problems involving the distribution of goods between warehouses and clients. These problems are known as VRPs. The VRP was introduced by Dantzig and Ramser, who called it the truck dispatching problem [DAN 54]. It involves planning the routes of a fleet of vehicles to serve a set of clients. Constructing the routes becomes more complex after constraints and objectives are added to the problem. Some of the most important examples include maximizing client satisfaction and minimizing the cost of the routes.

The selective vehicle routing problem (SVRP) is another problem derived from the VRP. The SVRP is one variant of a set of problems in a class called vehicle routing problems with profits (VRPPs) [FEI 01]. VRPPs attempt to construct one or several routes that maximize the total profit generated from clients, while minimizing the route length. By their nature, these problems are dual-objective problems.

5.5.1. *Problems similar to the VRP*

There are three classes of problems similar to the VRP [FEI 01] as follows:

– selective traveling salesman problem (STSP): the goal is to find a route that maximizes the total profit subject to an upper bound on the cost of the route;

– prize-collecting traveling salesman problem (PCTSP): the goal is to find a route of minimum cost subject to a lower bound on the total profit of the route;

– profitable tour problem (PTP): the goal is to find a route that optimizes a linear combination of the two previous objectives. Note that the notion of cost of a route often reduces to the duration or total length of the route, which is the sum of the travel times or lengths of the arcs that make up this route.

The orienteering problem (OP) is another well-known problem that is similar to the STSP.

5.5.2. *Mathematical formulation*

The earliest description of the STSP can be found in [GOL 87]. The authors consider a periodic vehicle routing problem that accounts for stock. A fleet of vehicles must deliver fuel to a large number of clients every day. The problem is modeled as an STSP where the profit measures the urgency. In the early 1980s, there was also a description of a similar problem, called the OP [TSI 84]. This problem was inspired by a sport practiced outdoors in mountainous or forested regions. Competitors with maps are challenged to travel from a starting point to a certain destination, within a fixed time period, while passing through certain checkpoints. A score is awarded for each checkpoint, and the goal is to maximize the total score. The STSP is in fact a special case of the OP where the starting point and the destination coincide. Some generalizations of the OP have also been discussed in the literature. Thus, Kantor and Rosenwein [KAN 92] presented the OP with time windows, where the checkpoints can only be visited within certain windows. In [CHA 96], the authors extended the problem to the case of a team consisting of several competitors.

Like the OP and the PTP, the STSP belongs to the class of NP-complete problems. Indeed, these problems are generalizations of the TSP, which is itself NP-complete. Below, we give an integer-valued formulation of the oriented version of the STSP. Let $G = (N, A)$ be a complete graph, where $N = \{1, \dots, n\}$ is the set of n vertices of the graph. Vertex 1 plays the role of the warehouse, and A is the set of edges. We assign a profit g_i to each vertex $i \in N$ (with $g_1 = 0$) and a cost c_{ij} to each edge $(i, j) \in A$ (the cost is often the length of the edge).

The STSP asks us to find an elementary cycle of maximum profit that contains vertex 1 and whose cost does not exceed some bound $C_{\max}$.

The formulation presented in [FEI 01] assigns a binary variable x_{ij} to each edge $(i, j) \in A$, where x_{ij} is equal to 1 if the corresponding edge is used in the solution and 0 otherwise. A binary variable y_i is also assigned to each vertex $i \in N$, where y_i is equal to 1 if the corresponding vertex is visited and 0 otherwise. Using these variables, the STSP can be formulated as the following integer program:

$$\max \sum_{i \in N} y_i p_i \quad [5.31]$$

[5.32]

$$\text{s.t.} \sum_{j \in N-\{i\}} x_{ij} = y_i, \ i \in N \quad [5.33]$$

[5.34]

$$\sum_{i \in N-\{j\}} x_{ij} = y_j, \ j \in N \quad [5.35]$$

[5.36]

$$\sum_{i \in N} \sum_{j \in N-\{i\}} C_{ij} x_{ij} \leq C_{\max} \quad [5.37]$$

[5.38]

$$x_{ij} \in \{0, 1\} \quad [5.39]$$

[5.40]

$$y_i \in \{0, 1\} \quad [5.41]$$

The constraints [5.33] and [5.35] are the assignment constraints. The constraint [5.37] imposes a maximum cost. The STSP can be solved using heuristics, metaheuristics or exact methods.

5.6. Using Matlab

Matlab documentation includes a (lengthy) script "TravellingSalesmanExample.m" on the TSP. Figure 5.6, generated by this script, is shown below.

As noted above, as an alternative, we could use a genetic algorithm to solve the problem. We will return to this type of algorithm in a subsequent volume to show how they can be used to solve a TSP.

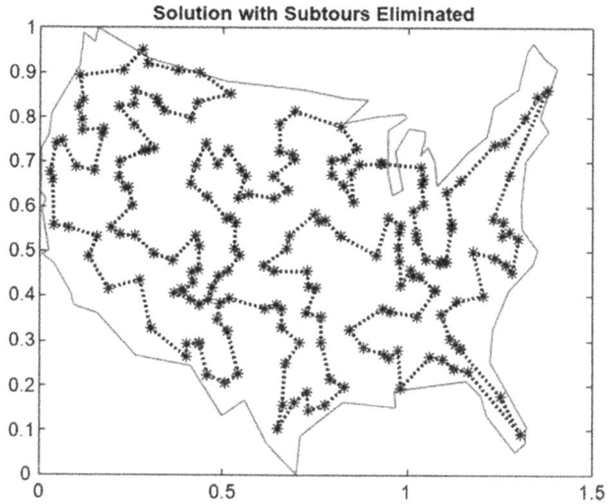

Figure 5.6. *Final tour of the traveling salesman. For a color version of this figure, see www.iste.co.uk/radi/optimizations.zip*

6

Unconstrained Nonlinear Programming

6.1. Introduction

Nonlinear programming (NLP) is primarily concerned with optimization problems whose assumptions (i.e. the functions and sets defining the problem) are nonlinear, but nevertheless sufficiently differentiable to apply theoretical tools such as optimality conditions, or solving algorithms specifically introduced and analyzed for this purpose. This subfield of optimization is somewhat ambiguously defined around the edges. Its introduction was arguably slightly artificial, and its existence is due to the community of researchers who specialize in these topics and the results that have been successfully established about them.

This chapter studies the problem of unconstrained nonlinear optimization, presenting its mathematical formulation and necessary optimality conditions. We then define a class of methods specifically designed for minimization problems. Finally, we end with an application of optimization techniques to identify the parameters of a structure under traction.

6.2. Mathematical formulation

Let $f\colon K \to \mathbb{R}$ be a function, with $K \subseteq \mathbb{R}^n$. The goal is to determine the vector $x \in K$ defined by:

$$f(x) = \min_{y \in K} f(y). \qquad [6.1]$$

Equivalently, we can find the value that maximizes f:

$$f(x) = \max_{y \in K} f(y). \qquad [6.2]$$

6.2.1. *Existence and uniqueness results*

The goal of this section is to specify a framework in which optimization problems admit at least one solution [ELM 10].

THEOREM 6.1.– Let $f\colon \mathbb{R}^n \to \mathbb{R}$ be a continuous and coercive function[1]. Then there is at least one element $\bar{x} \in \mathbb{R}^n$ such that

$$f(\bar{x}) \leq f(x) \quad \forall x \in \mathbb{R}^n.$$

Let $K \subset \mathbb{R}^n$ be a closed subset and $f\colon K \to \mathbb{R}$ a continuous function. Suppose further that:

$$\begin{cases} K \text{ bounded} \\ \text{or} \\ f(x) \to +\infty \text{ as } \|x\| \to \infty, \ x \in K \end{cases}$$

Then there exists at least one element $\bar{x} \in \mathbb{R}$ such that

$$f(\bar{x}) \leq f(x) \quad \forall x \in K.$$

THEOREM 6.2.– Let K be a convex set of $\mathbb{R}^n$, and let $f\colon K \to \mathbb{R}$ be a strictly convex function. Then the minimum $\bar{x} \in K$, if it exists, is unique.

REMARK.– In practice, (strict) convexity properties are rarely satisfied. If so, we can restrict to "local" aspects. We say there is a local minimum for $\bar{x}$ if there exists $\rho > 0$ such that

$$\|x - \bar{x}\| \leq \rho \Rightarrow f(\bar{x}) \leq f(x).$$

6.3. Optimality conditions

The following is restricted to the unconstrained case. Consider a given function $f\colon \mathbb{R}^n \to \mathbb{R}$ that is continuously differentiable. The optimization problem that we will consider is to find $\bar{x} \in \mathbb{R}^n$ such that

$$f(\bar{x}) \leq f(x) \quad \forall x \in \mathbb{R}^n. \qquad [6.3]$$

THEOREM 6.3.– A necessary condition for $\tilde{x}$ to be a solution of the optimization problem [6.3] is

$$\nabla f(\bar{x}) = 0. \qquad [6.4]$$

1 A function f is said to be coercive if $f(x) \to +\infty$ as $\|x\| \to \infty$.

REMARK.–

– The above theorem gives a necessary condition that is not sufficient in general.

– The necessary condition also holds for local minima.

THEOREM 6.4.– Let f be a convex function, and let $\bar{x} \in \mathbb{R}^n$ be such that $\nabla f(\bar{x}) = 0$. Then $\bar{x}$ is a solution of the global optimization problem [6.3].

EXAMPLE 6.1.– Let $A \in \mathcal{M}_{nn}(\mathbb{R})$ be a symmetric matrix, and let $b \in \mathbb{R}^n$. Consider the quadratic function f defined by $f(x) = \frac{1}{2}x^T Ax - b^T x$. The necessary optimality condition can be stated as

$$\nabla f(\bar{x}) = 0 \Leftrightarrow (A\bar{x} - b)^T = 0.$$

Hence, if A is a positive definite, the unique minimum is given by the equation

$$A\bar{x} = b.$$

The problem of minimizing a quadratic function can therefore be reduced to solving a linear system.

6.4. Quadratic problems

Let $A \in \mathcal{M}_{nn}(\mathbb{R})$ be a symmetric positive definite matrix, and let b be a vector of $\mathbb{R}^n$. Consider the quadratic form defined by [BER 01]:

$$f(x) = \frac{1}{2}x^T Ax - b^T x. \qquad [6.5]$$

Let d be an arbitrary direction of $\mathbb{R}^n$ ($d \neq 0$). Then, for any $t \in \mathbb{R}$:

$$f(x + td) = \frac{1}{2}(d^T Ad)t^2 + d^T(Ax - b)t + \left(\frac{1}{2}x^T Ax - b^T x\right).$$

Since A is a positive definite matrix, it follows that $d^T Ad > 0$. Thus, the function $t \mapsto f(x + td)$ is a parabola. The minimum is therefore attained for

$$\tilde{t} = -\frac{d^T(Ax - b)}{d^T Ad}. \qquad [6.6]$$

6.4.1. *Gradient method with optimal step size*

Here, we will consider the gradient method with an optimal step size. The direction of descent is $d = -\nabla f(x)^T = Ax - b$, so

$$\tilde{t} = \frac{||d||^2}{d^T Ad}, \qquad [6.7]$$

and the method can be stated in the form of [6.1].

Algorithm 6.1 Gradient method with optimal step size

Let $x_{(0)}$ be given.
Compute $d_{(0)} = b - Ax_{(0)}$.
For $k \geq 0$,

$$\begin{aligned} d_{(k)} &= b - Ax_{(k)}, \\ \alpha_k &= \frac{< d_{(k)}, d_{(k)} >}{< Ad_{(k)}, d_{(k)} >} \\ x_{(k+1)} &= x_{(k)} + \alpha_k d_{(k)} \end{aligned}$$

6.4.2. *Conjugate gradient method*

A method that is even faster in the case where A is a symmetric positive definite is the conjugate gradient method (Algorithm 6.2). This time, we set

$$x_{(k+1)} = x_{(k)} + \alpha_k d_{(k)}, \; k \geq 0, \qquad [6.8]$$

with $d_{(k)} = r_{(k)} - \beta_k d_{(k-1)}$.

Algorithm 6.2 Conjugate gradient method

Let $x_{(0)}$ be given.
Compute $r_{(0)} = b - Ax_{(0)}$, $d_{(0)} = r_{(0)}$.
For $k \geq 0$,

$$\begin{aligned} \alpha_k &= \frac{< r_{(k)}, d_{(k)} >}{< Ad_{(k)}, d_{(k)} >} \\ x_{(k+1)} &= x_{(k)} + \alpha_k d_{(k)} \\ r_{(k+1)} &= r_{(k)} - \alpha_k Ad_{(k)} \\ \beta_{k+1} &= \frac{< Ad_{(k)}, r_{(k+1)} >}{< Ad_{(k)}, d_{(k)} >} \\ d_{(k+1)} &= r_{(k+1)} - \beta_k d_{(k)}. \end{aligned}$$

6.5. Newton's algorithm

In optimization, Newton's algorithm can be applied directly to solve equations in the form $F(x) = 0$. In unconstrained optimization, this algorithm searches for the solutions of equation [6.4], i.e. the critical points of the function f being minimized. If we assume that f has class $\mathcal{C}^2$ and the Hessian matrix $H_f(x_k)$ is invertible, this algorithm can be iterated as follows:

$$x_{k+1} = x_k - H_f(x_k)^{-1}\nabla f(x_k), \qquad [6.9]$$

where $d_k = -H_f(x_k)^{-1}\nabla f(x_k)$ is called Newton's direction.

The direction d_k is also the unique solution of the problem:

$$\arg\min_{d\in\mathbb{R}^n} f(x_k) + \langle\nabla f(x_k),\ d\rangle + \frac{1}{2}\langle H_f(x_k)d, d\rangle. \qquad [6.10]$$

In other words, d_k is the global minimum point of the second-order approximation of f in the neighborhood of the current point x_k.

If the matrix $H_f(x_k)$ is positive definite at every iteration, we can view Newton's algorithm as a method of descent with a fixed step size of 1.

REMARK.–

1) It converges quadratically (the number of exact decimal places is multiplied by 2 at each iteration).

2) The difficulty lies in the cost of computing the Hessian $H_f(x_k)$: an analytic expression of the second derivatives is rarely available in applications.

3) This method does not converge if the first iteration is too far away from the solution, or if the Hessian is singular.

6.6. Methods of descent and linear search

6.6.1. *Presentation of methods of descent*

Let $f: \mathbb{R}^n \to \mathbb{R}$ be a differentiable function. Suppose that $x \in \mathbb{R}^n$ is also given.

The gradient vector $\nabla f(x)$ of f at x is a vector in $\mathbb{R}^n$ that is normal to the "surface" of isovalues[2] of f passing through x. It also points in the direction of increasing values of f.

DEFINITION 6.1.– *The vector $d \in \mathbb{R}^n$ is a direction of descent for f at the point x if*

$$\exists \tilde{t} > 0 / \ \forall t \in]0, \tilde{t}[,\ f(x + td) < f(x).$$

2 Recall that the isovalue sets of f are defined as $C_\alpha = \{x \in \mathbb{R}^n \mid f(x) = \alpha\}$ for $\alpha \in \mathbb{R}$.

EXAMPLE 6.2.– Let x be such that $\nabla f(x) \neq 0$. Suppose that $d \in \mathbb{R}^n$ is a direction such that

$$\nabla f(x).d < 0.$$

Then d is a direction of descent (see Figure 6.1).

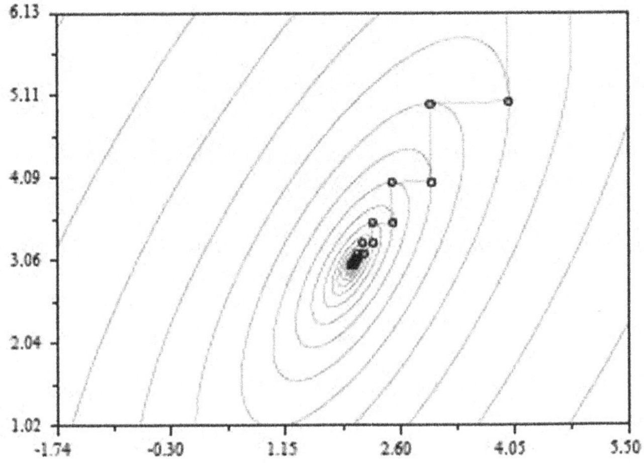

Figure 6.1. *Graphical solution. For a color version of this figure, see www.iste.co.uk/radi/optimizations.zip*

REMARK.– Thus, we can say that directions that form an acute angle with the opposite of the gradient are directions of descent. Indeed, we have

$$\begin{aligned}\nabla f(x).d < 0 &\Leftrightarrow -\nabla f(x).d > 0\\ &\leftrightarrow \theta \text{ acute,}\end{aligned}$$

where θ is defined by $\cos(\theta) = \dfrac{-\nabla f(x).d}{\|\nabla f(x).d\|}$.

Methods of descent are iterative methods. Their general procedure is given by Algorithm 6.3.

Choice of step size: The parameter α_k needs to be chosen as a function of the behavior of the function ϕ of one real variable defined by:

$$\phi(t) = f(x_{(k)} + \alpha d_k).$$

Algorithm 6.3 Method of descent

Let $x_{(0)} \in \mathbb{R}^n$ be given.
Construct a sequence $x_{(1)}, \ldots, x_{(k)}$ recursively:

$$x_{(k+1)} = x_{(k)} + \alpha_k d_k,$$

where d_k is a direction of descent at the point $x_{(k)}$ and $\alpha_k > 0$ is the step size of the method at the kth iteration.

For a direction of descent d_k, we know that $\phi(0) = f(x_{(k)})$ and $\phi'(0) = \nabla f(x_{(k)}).d_k < 0$. The optimal choice is therefore to choose t_k such that

$$\forall t > 0, \quad \phi(t_k) \leq \phi(t). \qquad [6.11]$$

In such a case, the method of descent is said to have optimal step size. However, calculating the optimal step size is a unidirectional optimization problem that is difficult to solve in general [NOC 99].

6.6.2. *Method of greatest slope*

Let $f : \mathbb{R}^n \to \mathbb{R}$ be a continuously differentiable function. Suppose that $x \in \mathbb{R}^n$ is given. In the neighborhood of x, the direction in which f decreases most rapidly is

$$d = -\nabla f(x).$$

In general, methods of greatest slope can be presented in the form:

$$\begin{cases} x_{(0)} \text{ given, } x_{(1)}, \ldots, x_{(k)} \text{ known :} \\ d_k = -\nabla f(x_{(k)})^T \\ \alpha_k = \arg\min_{\alpha \geq 0} f(x_k + \alpha d_k) \\ x_{(k+1)} = x_{(k)} + \alpha_k d_k \end{cases}$$

REMARK.– In the case of a gradient method with optimal step size, we write $\phi'(\alpha_k) = 0$ (necessary condition of optimality) by the inequality [6.11]. But

$$\phi'(\alpha_k) = \nabla f(x_{(k)} + \alpha_k d_k).d_k.$$

Hence,

$$\nabla f(x_{(k+1)}).d_k = 0.$$

and any two consecutive directions of descent are orthogonal.

6.6.3. *Acceptable step size*

In general, small step sizes should be used so that the linear approximation of the function given by the direction of greatest descent remains valid. In practice, it can be desirable to make progress more quickly by making larger steps. The choice of this step size generally follows two objectives that are often contradictory: finding the best step size and performing as few computations as possible.

Solving the problem

$$\alpha_k = \arg\min_{\alpha \geq 0} f(x_k + \alpha d_k) \qquad [6.12]$$

suggested by the algorithm can potentially be pointlessly expensive. Although this problem only has a single variable α, it is not trivial to solve in general. Furthermore, if the direction d_k computed by the algorithm is no good, it is not necessary to spend a lot of effort solving [6.12].

It can be shown that the condition $f(x_k + \alpha_k d_k) < f(x_k)$ is not sufficient for α_k to be considered as an acceptable step size. To avoid this inconvenience, we need to impose a condition to characterize the concept of a sufficient decrease in the function. To do this, we stipulate that a decrease that is deemed sufficient should be proportional to the step size. Thus, the larger the step size, the greater the decrease in the objective function. If $\gamma > 0$, we want

$$f(x_k) - f(x_k + \alpha_k d_k) \geq \alpha_k \gamma.$$

The factor γ cannot be chosen arbitrarily: it must vary from one direction to another. It is more appropriate to define γ as a function of the slope of the function at the point x_k in the direction d_k. Let us choose:

$$\gamma = -\beta \nabla f(x_k)^T d_k,$$

with $0 < \beta < 1$. Thus, as the directional derivative $\nabla f(x_k)^T d_k$ approaches zero, the slope of the line becomes smaller and vice versa.

DEFINITION 6.2.– *Let* $f\colon \mathbb{R}^n \to \mathbb{R}$ *be a differentiable function,* $x_k \in \mathbb{R}^n$ *a point,* $d_k \in \mathbb{R}^n$ *a direction of descent such that* $\nabla f(x_k)^T d_k < 0$ *and* $\alpha_k \in \mathbb{R}$ *a step size,* $\alpha > 0$. *We say that the function* f *decreases sufficiently at* $x_k + \alpha_k d_k$ *with respect to* x_k *if*

$$f(x_k + \alpha_k d_k) \leq f(x_k) + \alpha_k \beta_1 \nabla f(x_k)^T d_k, \qquad [6.13]$$

with $0 < \beta_1 < 1$.

The condition [6.13] is called Wolfe's first condition, after Wolfe (1969), or Armijo's condition, after Armijo (1966). It allows us to reject step sizes that are too large and do not achieve a sufficient decrease in the function. Now that we have solved the problem of step sizes that are too large, let us turn our attention to step sizes that are too small.

DEFINITION 6.3.– *Let* $f\colon \mathbb{R}^n \to \mathbb{R}$ *be a differentiable function,* $x_k \in \mathbb{R}^n$ *a point,* $d_k \in \mathbb{R}^n$ *a direction of descent such that* $\nabla f(x_k)^T d_k < 0$ *and* $\alpha_k \in \mathbb{R}$ *a step size,* $\alpha > 0$*. We say that the point* $x_k + \alpha_k d_k$ *achieves sufficient progress with respect to* x_k *if*

$$\nabla f(x_k + \alpha_k d_k)^T d_k \geq \beta_2 \nabla f(x_k)^T d_k \qquad [6.14]$$

with $0 < \beta_2 < 1$. The condition [6.14] is called Wolfe's second condition. We have the following result:

THEOREM 6.5.– Let $f\colon \mathbb{R}^n \to \mathbb{R}$ be a differentiable function, $x_k \in \mathbb{R}^n$ a point and $d_k \in \mathbb{R}^n$ a direction of descent such that $\nabla f(x_k)^T d_k < 0$. Suppose that f is bounded from below in the direction d_k, i.e. there exists f_0 such that $f(x_k + \alpha d_k) \geq f_0$ for every $\alpha \geq 0$. If $0 < \beta_1 < 0$, there exists η such that Wolfe's first condition [6.13] is satisfied for every $\alpha_k \leq \eta$. Furthermore, if $0 < \beta_1 < \beta_2 < 1$, there exists $\alpha_2 > 0$ such that Wolfe's two conditions [6.13] and [6.14] are both satisfied.

6.6.4. *Linear search*

Linear search allows us to identify a step size that satisfies Wolfe's conditions [6.13] and [6.14]. The first linear search algorithm according to Wolfe's conditions, attributed to Fletcher [FLE 80] and Lemaréchal [LEM 81], is presented in Algorithm 6.4.

We have the following convergence result:

THEOREM 6.6.– Let $f\colon \mathbb{R}^n \to \mathbb{R}$ be a differentiable function. Let $x_k \in \mathbb{R}^n$ and d_k be a direction of descent of f at x_k. Suppose that f is bounded from below in the direction d_k. Then Wolfe's linear search algorithm terminates in finitely many iterations.

There is also a general result about global convergence for descent algorithms with Wolfe step sizes. This result, stated below, involves the angle θ_k between the direction of descent d_k and the direction of the gradient, defined by:

$$\cos(\theta_k) = \frac{\langle \nabla f(x_k), d_k \rangle}{\| \nabla f(x_k) \| \| d_k \|}. \qquad [6.15]$$

We have the following result:

THEOREM 6.7.– Let $f\colon \mathbb{R}^n \to \mathbb{R}$ be differentiable with a Lipschitz gradient and bounded from below. Let A be an algorithm that generates iterations defined by:

$$x_{k+1} = x_k + \alpha_k d_k,$$

where d_k is a direction of descent of f at x_k and $\alpha_k > 0$ is a step size satisfying Wolfe's conditions. Then $\sum \cos(\theta_k)^2 \parallel \nabla f(x_k) \parallel^2$ converges.

Algorithm 6.4 Wolfe's linear search

1: **Assumptions**: f function of class $\mathcal{C}^1$, $x \in \mathbb{R}^n$ current point, d direction of descent of f at x, α_0 initial approximation of the solution, $\epsilon_1 > 0$ and $\epsilon_2 > 0$ such that: $0 < \epsilon_1 < \epsilon_2 < 1$.

2: $k \leftarrow 0\,;\, \alpha_- = 0\,;\, \alpha_+ = +\infty$.

3: **While** α_k does not satisfy Wolfe's two conditions,

a) If α_k does not satisfy [6.13], then the step size is too large and:

$$\alpha_+ = \alpha_k \quad \text{and} \quad \alpha_{k+1} = \frac{\alpha_- + \alpha_+}{2}.$$

b) If α_k satisfies [6.13] but not [6.14], then the step size is too small and:

$$\alpha_- = \alpha_k \quad \text{and} \quad \alpha_{k+1} = \begin{cases} \frac{\alpha_- + \alpha_+}{2} & \text{if } \alpha_+ < +\infty \\ 2\alpha_k & \text{otherwise.} \end{cases}$$

c) $k \leftarrow k + 1$.

4: **Return** α_k.

5: **Output:** a step size α^* satisfying Wolfe's conditions.

6.6.5. *Newton's method with linear search*

Consider a very general Newton-type iteration:

$$x_{k+1} = x_k - s_k H_k^{-1} \nabla f(x_k). \qquad [6.16]$$

We want to require H_k to be positive definite to guarantee that $-H_k^{-1}\nabla f(x_k)$ is indeed a direction of descent of f at x_k.

If the matrix $H[f](x_k)$ is positive definite, then: $H_k = H[f](x_k)$. If Newton's step size ($s_k = 1$) is acceptable in the sense of Wolfe's conditions, then the algorithm performs a Newton iteration; if not, it performs a Wolfe linear search initialized with a step size of 1.

If not, the most widely used technique is to generate a matrix E such that $H_k = H[f](x_k) + E$ is positive definite. Note that this is always possible if we choose E of the form αI.

REMARK.– In general, algorithms compute a Cholesky factorization $L_k L_k^T$ of $H[f](x_k) + \alpha I$, where the parameter $\alpha > 0$ is increased until the factorization becomes possible. The parameter α is initialized as follows: $\alpha_0 = 0$ if $H[f](x_k)$ is positive definite, $\alpha_0 = \frac{1}{2} \parallel H[f](x_k) \parallel_F^2$ otherwise.

We then solve:

$$L_k z_k = \nabla f(x_k) \quad \text{and} \quad L_k^T d_k = -z_k.$$

The advantage is that the Hessian becomes a positive definite at the current point, and so, we recover Newton's algorithm and its quadratic convergence.

Suppose that f is of class C^1 with a Lipschitz gradient and is bounded from below. The choice of the direction of descent in this algorithm is as follows: $d_k = -H_k^{-1}\nabla f(x_k)$, where H_k is a symmetric positive definite matrix. When coupled with a Wolfe linear search, we have the following result:

THEOREM 6.8.– If the conditioning of the matrix H_k remains bounded over the iterations, i.e. if:

$$\exists M > 0, \ \forall k \in \mathbb{N}, \ \kappa_2(H_k) = \parallel H_k \parallel_2 \parallel H_k^{-1} \parallel_2 \leq M,$$

then Newton's algorithm with a Wolfe step size converges globally.

6.7. Quasi-Newton methods

Coupling Newton's method with a Wolfe linear search allowed us to construct a globally convergent method. Quasi-Newton methods were developed to circumvent other disadvantages of Newton's method: in particular the problem of computing the Hessian matrix, which is not always possible or advisable. These methods instead focus on iteratively constructing matrices H_k that approach the Hessian, or matrices B_k that approach the inverse of the Hessian.

How can we compute an approximation H_{k+1} of the Hessian matrix $H[f](x_{k+1})$ if we know x_k and x_{k+1}, as well as $\nabla f(x_k)$ and $\nabla f(x_{k+1})$?

Let us write the series expansion of ∇f in the neighborhood of x_{k+1} at x_k:

$$\nabla f(x_k) = \nabla f(x_{k+1}) + H[f](x_{k+1})(x_k - x_{k+1}) + o(x_k - x_{k+1}). \quad [6.17]$$

Hence:

$$H[f](x_{k+1})(x_k - x_{k+1}) \approx \nabla f(x_{k+1}) - \nabla f(x_k). \quad [6.18]$$

We can now construct an approximation H_{k+1} of $H[f](x_{k+1})$ as a solution of the equation:

$$\nabla f(x_{k+1}) - \nabla f(x_k) = H_{k+1}(x_{k+1} - x_k), \quad [6.19]$$

called the secant equation or quasi-Newton equation. Similarly, we can construct an approximation B_{k+1} of $H[f](x_{k+1})^{-1}$ as a solution of the equation:

$$B_{k+1}(\nabla f(x_{k+1}) - \nabla f(x_k)) = x_{k+1} - x_k. \qquad [6.20]$$

In both cases, the quasi-Newton equations form an underdetermined system of n equations and n^2 unknowns. There are therefore infinitely many potentially suitable matrices H_{k+1}.

6.7.1. *DFP and BFGS methods*

From the infinitely many possible linear models, Broyden [BRO 65] suggested choosing the model that is closest to the model established in the previous iteration in order to preserve as many of the previous computations as possible. We therefore wish to solve:

$$\begin{cases} \min_H \frac{1}{2} \| H - H_k \|_F^2 \\ \text{s.t.} \quad \nabla f(x_{k+1}) - \nabla f(x_k) = H(x_{k+1} - x_k) \end{cases} \qquad [6.21]$$

where $\| \,.\, \|_F$ denotes the Frobenius norm. This is a strictly convex quadratic problem whose unique solution is given by:

$$H_{k+1} = H_k + \frac{(y_k - H_k\sigma_k)\sigma_k^T}{\sigma_k^T\sigma_k}, \qquad [6.22]$$

where:

$$\begin{cases} \sigma_k = x_{k+1} - x_k \\ y_k = \nabla f(x_{k+1}) - \nabla f(x_k). \end{cases} \qquad [6.23]$$

The problem with this method is that the matrix H_k that we have constructed is neither a symmetric nor positive definite in general. Since the Hessian of a function of class C^2 is symmetric, it is natural to require its approximations to also be symmetric. We therefore want to find H_{k+1} as a solution of the problem:

$$(P) \begin{cases} \min_H \frac{1}{2} \| H - H_k \|^2 \\ \text{s.t.} \quad y_k = H\sigma_k, \; H^T = H. \end{cases} \qquad [6.24]$$

Equation matrix norms can be used, each leading to a different quasi-Newton method. The most widely used methods are the DFP and BFGS methods obtained by choosing norms of the form:

$$\| A \|_W = \| W^{\frac{1}{2}} A W^{\frac{1}{2}} \|_F, \qquad [6.25]$$

where W is an invertible symmetric matrix satisfying $y_k = W\sigma_k$. Solving the problem (P) leads to the following update formulas:

– DFP method (Davidson, Fletcher, Powell)

$$H_{k+1} = H_k + \frac{y_k y_k^T}{y_k^T \sigma_k} - \frac{H_k \sigma_k \sigma_k^T H_k}{\sigma_k^T H_k \sigma_k}. \qquad [6.26]$$

– BFGS method (Broyden, Fletcher, Goldfarb, Shannon)

$$B_{k+1} = \left(I - \frac{\sigma_k y_k^T}{y_k^T \sigma_k}\right)^T B_k \left(I - \frac{\sigma_k y_k^T}{y_k^T \sigma_k}\right) + \frac{\sigma_k \sigma_k^T}{y_k^T \sigma_k}. \qquad [6.27]$$

These two formulas are dual to one another in the sense that the BFGS formula is obtained by inverting the DFP relation: $H_k = B_k^{-1}$. We also have $H_{k+1} = B_{k+1}^{-1}$.

We have the following result:

THEOREM 6.9.– If B_k is a positive definite and $y_k^T \sigma_k > 0$, then B_{k+1} is a positive definite.

As above, the BFGS algorithm will be a descent algorithm whenever it is initialized with a symmetric positive definite matrix H_0 while imposing the following condition on each iteration:

$$(\nabla f(x_{k+1}) - \nabla f(x_k))^T (x_{k+1} - x_k) > 0. \qquad [6.28]$$

Note that this inequality is satisfied whenever a Wolfe linear search is used to compute the step size.

REMARK.– BFGS is considered to be a very effective algorithm. Although the algorithm does not have a convergence result in the non-convex case, there is nothing to prevent us from applying it anyway: the iterations are well defined and the algorithm often achieves remarkable numerical results.

6.8. Relaxation method

This method makes the following choice for the directions of descent [BON 06]:

$$d^{(0)} = e_1,\ d^{(1)} = e_2, \ldots, d^{(n-1)} = e_n,$$

then, starting again:

$$d^{(n)} = e_1,\ d^{(n+1)} = e_2, \ldots,\ d^{(2n-1)} = e_n,$$

and so on, where $\{e_1, e_2, \ldots, e_n\}$ are the vectors of the canonical basis of $\mathbb{R}^n$. Therefore, in general, we have $d^{(k)} = e_l$ if and only if l is the remainder of $k+1$ divided by n.

Algorithm 6.5 BFGS algorithm

Assumptions: $f\colon \mathbb{R}^n \to \mathbb{R}$, x_0 arbitrary initial point and H_0 arbitrary symmetric positive definite matrix.

$k \leftarrow 0$

While stopping test not satisfied,

1) choose the direction: $d_k = -H_k^{-1}\nabla f(x_k)$;

2) Wolfe linear search to compute a step size $\alpha_k > 0$.

3) New iteration:

$$x_{k+1} = x_k - \alpha_k H_k^{-1}\nabla f(x_k).$$

4) Update the matrix H_k at H_{k+1} with the BFGS formula.

$k \leftarrow k+1$.

Return x_k.

Output: an approximation of the solution of the problem: $\min\limits_{x_o\in\mathbb{R}^n} f(x)$.

Next, we take the factors $\rho_k \in \mathbb{R}$ such that

$$J(u^{(k)} + \rho_k d^{(k)}) = \min_{\rho\in\mathbb{R}} J(u^{(k)} + \rho d^{(k)}),\ \forall k \in \mathbb{N} \qquad [6.29]$$

(assuming that these minima exist).

Finally, we set

$$u^{(k+1)} = u^{(k)} + \rho_k d^{(k)}.$$

This is the relaxation method (or algorithm).

Equivalently, we can rewrite this algorithm as follows:

In the above, we used the fact that

$$\min_{\rho\in\mathbb{R}} J(u_1^{(k)} + \rho, u_2^{(k)}, \dots, u_n^{(k)}) = \min_{y\in\mathbb{R}} J(y, u_2^{(k)}, \dots, u_n^{(k)}).$$

This equality is obtained by performing the change of variable $y = u_1^{(k)} + \rho$.

The key theoretical result of this section is the following:

THEOREM 6.10.– Let $J\colon \mathbb{R}^n \to \mathbb{R}$ be an elliptic function. Then the relaxation method is well defined and converges (i.e. for every $u^{(0)} \in \mathbb{R}^n$, the sequence $u^{(k)}$ constructed by this algorithm converges to the unique minimum point of J).

Algorithm 6.6 Relaxation algorithm

Suppose that the vector $u^{(k)} = (u_1^{(k)}, u_2^{(k)}, \ldots, u_n^{(k)})^T$ is known.
Compute $u^{(k+1)} = (u_1^{(k+1)}, u_2^{(k+1)}, \ldots, u_n^{(k+1)})^T$ in n successive steps with the following formulas:

$$J(u_1^{(k+1)}, u_2^{(k)}, \ldots, u_n^{(k)}) = \min_{y \in \mathbb{R}} J(y, u_2^{(k)}, \ldots, u_n^{(k)})$$

$$J(u_1^{(k+1)}, u_2^{(k+1)}, u_3^{(k)}, \ldots, u_n^{(k)}) = \min_{y \in \mathbb{R}} J(u_1^{(k+1)}, y, u_3^{(k)}, \ldots, u_n^{(k)})$$

$$\vdots$$

$$J(u_1^{(k+1)}, \ldots, u_n^{(k+1)}) = \min_{y \in \mathbb{R}} J(u_1^{(k+1)}, \ldots, u_{n-1}^{(k+1)}, y)$$

6.9. Gradient method

Suppose that the function $f\colon \mathbb{R}^n \to \mathbb{R}$ is differentiable. A gradient-type algorithm takes the opposite of the direction of the gradient of f at the current point as the direction of descent:

$$x_0 \in \mathbb{R}^n \text{ arbitrary, } x_{k+1} = x_k - \alpha_k \nabla f(x_k), \qquad [6.30]$$

where the step size $\alpha_k > 0$ is determined by a linear search strategy. Recall that $d_k = -\nabla f(x_k)$ is indeed a direction of descent of f at x_k and that it is the direction of the greatest slope of f at x_k.

We want a sufficient condition on the step size α_k to guarantee that Algorithm 6.30 is a method of descent, i.e. $\forall k \in \mathbb{N}$, $f(x_{k+1}) < f(x_k)$.

THEOREM 6.11.– Let f be a function bounded from below class C^1 with a Lipschitz gradient with Lipschitz constant $L > 0$. If $\alpha_k < \frac{2}{L}$ for every k, then the algorithm: $x_0 \in \mathbb{R}^n$ arbitrary, $x_{k+1} = x_k - \alpha_k \nabla f(x_k)$ is indeed a descent algorithm, i.e. $\forall k \in \mathbb{N}$, $f(x_{k+1}) < f(x_k)$.

Suppose now that the step size α_k does indeed satisfy $\alpha_k < \frac{2}{L}$ for every $k \in \mathbb{N}$. We have the following global convergence results:

THEOREM 6.12.– Let f be a function bounded from below class C^1 with a Lipschitz gradient with Lipschitz constant $L > 0$. Then:

1) If, for every k, $\alpha_k := \alpha < \frac{2}{L}$, then the gradient descent algorithm with a constant step size converges globally and:

$$0 < \alpha\left(1 - \frac{L}{2}\alpha\right) \| \nabla f(x_k) \|^2 \leq f(x_k) - f(x_{k+1}).$$

Furthermore, the constant step size that yields the best possible decrease is $\alpha_k = \frac{1}{L}$.

2) The gradient algorithm with an optimal step size converges globally and

$$\frac{1}{2L} \parallel \nabla f(x_k) \parallel^2 \leq f(x_k) - f(x_{k+1}).$$

3) The gradient algorithm with a Wolfe step size converges globally.

6.10. Least squares problem

Suppose that the function F is linear and can be written in the form $F(x) = Ax - b$, with $A \in \mathcal{M}_{n,p}(\mathbb{R})$. Let us solve the following linear least squares problem:

$$\min_{x \in \mathbb{R}^n} r(x) = \frac{1}{2} \parallel Ax - b \parallel_2^2 . \qquad [6.31]$$

We compute the Jacobian of F at each point: $J_F(x) = A$, then deduce:

$$\nabla r(x) = A^T(Ax - b) \quad \text{and} \quad H_r(x) = J_F(x)^T J_F(x) = A^T A. \qquad [6.32]$$

The Hessian of r is positive at every point, so the least squares problem is convex. We need to find the critical points of the problem [6.32], i.e. the points that satisfy $\nabla r(x) = 0$, which gives us the following system of normal equations to solve:

$$A^T Ax = A^T b. \qquad [6.33]$$

Thus, x^* is a solution of the problem [6.32] if and only if $A^T Ax^* = A^T b$. Furthermore, if A is full-rank, then the function r is strictly convex and x^* is the unique solution of [6.32].

Now, if we apply Newton's method to the problem [6.32], Newton's direction $d_k = x_{k+1} - x_k$ is a solution of the system $A^T Ad_k = -A^T(Ax_k - b)$, so $A^T Ax_{k+1} = A^T b$, and this holds for any x_k. This can be recognized as the system of normal equations from the problem [6.32].

6.10.1. *Gauss–Newton method*

Let us now consider the general problem of nonlinear least squares [6.31]. There are two ways to present Newton's algorithm:

1) the first is to apply Newton's method to the problem [6.31] while neglecting the second-order term $\sum_{i=1}^m F_i(x)H_{F_i}(x)$ in the Hessian of r;

2) the second is to replace the nonlinear least squares problem by an approximate problem of linear least squares at each iteration.

Let $x_k \in \mathbb{R}^n$ be the current point. In the neighborhood of x_k, we replace the problem [6.31] by:

$$\min_{y \in \mathbb{R}^n} \tilde{r}(y) = \frac{1}{2} \parallel F(x_k) + J_F(x_k)(y - x_k) \parallel_2^2, \qquad [6.34]$$

where we have replaced F by a first-order approximation at x_k. The problem [6.34] is a linear least squares problem whose solutions x_{k+1} satisfy the corresponding normal equations, namely:

$$J_F(x_k)^T J_F(x_k)(x_{k+1} - x_k) = -J_F(x_k)^T F(x_k). \qquad [6.35]$$

The direction $d_k = x_{k+1} - x_k$ that this defines is called the Gauss–Newton direction.

Algorithm 6.7 Gauss–Newton algorithm

1: **Assumptions**: F differentiable function, x_0 initial point, $\epsilon > 0$ required precision.
2: $k \leftarrow 0$.
3: **While** stopping criterion that remains to be defined,

1) compute a search direction: compute the solution d_k of:

$$J_F(x_k)^T J_F(x_k) d = -J_F(x_k)^T F(x_k).$$

2) $x_{k+1} = x_k + d_k$.

4: $k \leftarrow k + 1$.
5: **Return** x_k.
6: **Output**: an approximation of the least squares problem:

$$\min_{x \in \mathbb{R}^n} r(x) = \frac{1}{2} F(x)^T F(x).$$

The speed of convergence of the Gauss–Newton algorithm depends on the quality of the approximation of the initial problem. In the case where the residual is effectively small in the neighborhood of the solution, the convergence occurs quickly. If the Hessian of F is zero at the solution, the convergence is quadratic.

REMARK.– This method has the following advantages compared to Newton's method:

1) the approximation means that we do not need to compute second derivatives;

2) in various applications, the approximation is of good quality;

3) if $J_F(x_k)$ is full-rank and $\nabla r(x_k)$ is non-zero, then the Gauss–Newton direction is a direction of descent of r at x_k and a linear search algorithm can therefore be used.

With additional hypotheses, we can show that the iterations converge globally to a stationary point r. One disadvantage of the Gauss–Newton method is that we do not have a convergence result when the Jacobian of F at x_k is not full-rank. To circumvent this issue, we can use the Levenberg–Marquardt method.

6.10.2. *Levenberg–Marquardt algorithm*

This algorithm is a regularization of the Gauss–Newton algorithm that can be used when the Jacobian of F is not full-rank, among other things. Nevertheless, it has a close connection to the so-called trust region methods. This is the perspective taken to describe the algorithm here.

Consider the current iteration x_k. Like for the Gauss–Newton method, we replace the nonlinear least squares problem [6.32] by the following problem in the neighborhood of x_k:

$$\min_{y \in \mathbb{R}^n} \frac{1}{2} \| F(x_k) + J_F(x_k)(y - x_k) \|_2^2,$$

this time also adding the constraint $\| y - x_k \|^2 \leq \Delta_k$. The parameter $\Delta_k > 0$ is called the radius of the trust region. At each iteration, the initial problem is replaced by the following quadratic problem:

$$(P_k) \begin{cases} \min_{y \in \mathbb{R}^n} \frac{1}{2} \| F(x_k) + J_F(x_k)(y - x_k) \|_2^2 \\ \text{s.t.} \quad \| y - x_k \|_2 \leq \Delta_k. \end{cases} \qquad [6.36]$$

By introducing the Lagrangian corresponding to the problem (P_k):

$$\mathcal{L}_k(y, \lambda) = \frac{1}{2} \| F(x_k) + J_F(x_k)(y - x_k) \|_2^2 + \frac{1}{2}\lambda(\| y - x_k \|^2 - \Delta_k^2), \ \lambda \in \mathbb{R}, \qquad [6.37]$$

we can state the first-order necessary conditions for optimality as follows:

$$\begin{aligned} &(J_F(x_k)^T J_F(x_k) + \lambda I)d = -J_F(x_k)^T F(x_k) \\ &\lambda = 0 \text{ or } \| y - x_k \| = \Delta_k \\ &\lambda = 0 \end{aligned} \qquad [6.38]$$

In the case where the trust region constraint is inactive (i.e. $\| y - x_k \| < \Delta_k$ and $\lambda = 0$), this reduces to an iteration of the Gauss–Newton algorithm. In the case where the constraint ($\| y - x_k \| = \Delta_k$) is active, the solution is on the boundary of the trust region. The step size can be seen as a regularization of the Gauss–Newton step size in the sense that the matrix $(J_F(x_k)^T J_F(x_k) + \lambda I)$ is now a positive definite: the step size therefore corresponds to a descent of the function r at x_k.

Algorithm 6.8 Levenberg–Marquardt algorithm

1: **Assumptions**: F differentiable function, x_0 initial point, $\epsilon > 0$ required precision.
2: $k \leftarrow 0$.
3: **While** stopping criterion that remains to be defined,

1) Compute a search direction: compute the solution d_k of:

$$(J_F(x_k)^T J_F(x_k) + \lambda I)d = -J_F(x_k)^T F(x_k).$$

2) $x_{k+1} = x_k + d_k$.

3) Update the parameter λ.

4: $k \leftarrow k + 1$.
5: **Return** x_k.
6: **Output**: an approximation of the solution of the least squares problem:

$$\min_{x \in \mathbb{R}^n} r(x) = \frac{1}{2} F(x)^T F(x).$$

REMARK.– The parameter $\lambda \geq 0$ can be fixed or adjusted heuristically by increasing it or decreasing by a factor according to the quality of the proposed step size.

6.10.3. *Kalman filter*

When calibrating mathematical models based on observations of a real system, the data is often structured into blocks. In the linear case, suppose that the least squares problem

$$\min_{x \in \mathbb{R}^n} \|Ax - b\|_2^2$$

can be decomposed into J blocks, each containing m_j pieces of data. The problem can then be written as

$$\min_{x \in \mathbb{R}^n} \sum_{j=1}^{J} \|A_j x - b_j\|_2^2,$$

where $A_j \in \mathbb{R}^{m_j \times n}$ and $b_j \in \mathbb{R}^{m_j}$.

We will construct the solution of this problem incrementally, starting by solving the subproblem corresponding to the first block of data:

$$\min_{x \in \mathbb{R}^n} \|A_1 x - b_1\|_2^2.$$

Suppose that the matrix A_1 is full-rank. In this case, the solution x_1 is given by the normal equations:

$$A_1^T A_1 x_1 = A_1^T b_1. \qquad [6.39]$$

Since A_1 is full-rank, $A_1^T A_1$ is invertible and

$$x_1 = (A_1^T A_1)^{-1} A_1^T b_1. \qquad [6.40]$$

We can construct an incremental form starting from any x_0:

$$\begin{aligned} x_1 &= x_0 - x_0 + (A_1^T A_1)^{-1} A_1^T b_1 \\ &= x_0 - (A_1^T A_1)^{-1} A_1^T A_1 x_0 + (A_1^T A_1)^{-1} A_1^T b_1 \\ &= x_0 + (A_1^T A_1)^{-1} A_1^T (b_1 - A_1 x_0). \end{aligned}$$

Now consider the problem consisting of the first two blocks:

$$\min_{x \in \mathbb{R}^n} \|A_1 x - b_1\|_2^2 + \|A_2 x - b_2\|_2^2.$$

which we can also rewrite as

$$\min_{x \in \mathbb{R}^n} \| \begin{bmatrix} A_1 \\ A_2 \end{bmatrix} x - \begin{bmatrix} b_1 \\ b_2 \end{bmatrix} \|_2^2$$

whose normal equations are

$$(A_1^T A_1 + A_2^T A_2) x_2 = A_1^T b_1 + A_2^T b_2.$$

By [6.39], we obtain

$$\begin{aligned} (A_1^T A_1 + A_2^T A_2) x_2 &= A_1^T b_1 + A_2^T b_2 \\ &= A_1^T A_1 x_1 + A_2^T b_2 \\ &= A_1^T A_1 x_1 + A_2^T A_2 x_1 - A_2^T A_2 x_1 + A_2^T b_2 \\ &= (A_1^T A_1 + A_2^T A_2) x_1 + A_2^T (b_2 - A_2 x_1) \end{aligned}$$

Since $A_1^T A_1$ is invertible, $A_1^T A_1 + A_2^T A_2$ is also invertible, and we obtain:

$$x_2 = x_1 + (A_1^T A_1 + A_2^T A_2)^{-1} A_2^T (b_2 - A_2 x_1).$$

Repeating the same technique for the next blocks gives the Kalman filter algorithm.

Algorithm 6.9 Kalman filter algorithm

1: **Inputs:**
- The matrices $A_j \in \mathbb{R}^{m_j \times n}$, $j = 1, \dots, J$.
- The vectors $b_j \in \mathbb{R}^{m_j}$, $j = 1, \dots, J$.
- An initial solution $x_0 \in \mathbb{R}^n$.
- An initial filter $H_0 \in \mathbb{R}^{n \times n}$.

2: **Initialization:** $j = 1$.
3: **Iterations:**
4: $H_j = H_{j-1} + A_j^T A_j$.
5: $x_j = x_{j-1} + H_j^{-1} A_j^T (b_j - A_j x_{j-1})$.
6: **Stopping criterion:** When $j = J$, $x^* = x_J$.
7: **Output:** The solution $x^* \in \mathbb{R}^n$.

6.11. Direct search methods

We can cite two families of methods that do not involve calculating the derivative:

– interpolant methods;

– direct search:

1) Nelder–Mead;

2) Torczon.

6.11.1. *Nelder–Mead algorithm*

The Nelder–Mead algorithm is a nonlinear optimization algorithm published by Nelder and Mead in 1965 [NEL 65]. It is a heuristic numerical method that aims to find (an approximation of) a local minimum of the problem of minimizing a continuous function over a multidimensional space. It is sometimes called the simplex method, but should not be confused with the simplex algorithm from linear programming.

A k-dimensional simplex is the convex envelope of $k + 1$ affinely independent vectors $x_1, \dots, x_{k+1}$ of $\mathbb{R}^n$, $k \leq n$, i.e. the k vectors $x_1 - x_{k+1}, x_2 - x_{k+1}, \dots, x_k - x_{k+1}$ are linearly independent.

EXAMPLE 6.3.– Any three non-aligned points in $\mathbb{R}^2$ or any four non-coplanar points in $\mathbb{R}^3$ are affinely independent, defining a two- and three-dimensional simplex, respectively.

Let f be a function defined on an N-dimensional space. The algorithm starts by defining a non-degenerate simplex in this space. By successive iterations, the process determines the point of this simplex where the function is maximal, then replaces this point with the point symmetric to it with respect to the center of gravity of the N remaining points. If the value of the function at this new point is lower than every other value on the other points, the simplex is stretched in this direction. Otherwise, if the value of the function at this new point is better than the second-best but worse than the best, the value is kept, and we start again. If not, we assume that the function is locally shaped like a valley, and the simplex is contracted onto itself. If this still does not give a better point, the simplex is reduced by a homothety centered at the point of the simplex where the function is minimal.

Algorithm 6.10 Nelder–Mead algorithm

Inputs:
- The continuously differentiable function $f\colon \mathbb{R}^n \to \mathbb{R}$.
- The affinely independent vectors $x_1^0, \dots, x_{n+1}^0$ of $\mathbb{R}^n$ such that $f(x_i^0) \leq f(x_{i+1}^0)$, $i = 1, \dots, n$,, forming a simplex.
- The required precision $\epsilon \in \mathbb{R}_+^*$.

Initialization: $k \leftarrow 0$

Iterations:

Compute the center of gravity of the other points of the simplex $x_c = \frac{1}{n}\sum_{i=1}^{n} x_i^k$.

Define the direction $d_k = x_c - x_{n+1}^k$.

Compute $x_r = x_{n+1}^k + 2d_k = 2x_c - x_{n+1}^k$.

If $f(x_r) < f(x_1^k)$, then search beyond x_r:
a) Compute $x_e = x_{n+1}^k + 3d_k = 2x_r - x_c$.
b) If $f(x_e) < f(x_r)$, then $\bar{x} = x_e$, else $\bar{x} = x_r$.

If $f(x_n^k) > f(x_r) \geq f(x_1^k)$, then $\bar{x} = x_r$.

If $f(x_r) \geq f(x_n^k)$, then search before x_r:
a) If $f(x_r) \geq f(x_{n+1}^k)$, then $\bar{x} = x_{n+1}^k + \frac{1}{2}d_k = \frac{1}{2}(x_{n+1}^k + x_c)$.
b) If $f(x_r) < f(x_{n+1}^k)$, then $\bar{x} = x_{n+1}^k + \frac{3}{2}d_k = \frac{1}{2}(x_r + x_c)$.

$x_{n+1}^{k+1}\bar{x}, x_i^{k+1} = x_i^k, i = 1, \dots, n.$

$k \leftarrow k + 1.$

Renumber to obtain $f(x_i^k) \leq f(x_{i+1}^k)$, $i = 1, \dots, n$.

Stopping criterion: If $\|d_k\| \leq \epsilon$, then $x^* = x_1^k$.

Output: An approximation of the solution $x^* \in \mathbb{R}$.

6.11.2. *Torczon method*

One of the issues with the Nelder–Mead method is the risk that the successively constructed simplexes might be degenerate, or that three points might be almost colinear. The Torczon method, also known as Torczon multidirectional search, maintains the geometry of the simplex and does not encounter problems with degeneracy in general. It follows the same objective as the Nelder–Mead algorithm. The details of this method are presented in Algorithm 6.11.

Algorithm 6.11 Torczon algorithm

1: **Inputs**
- The function $f\colon \mathbb{R}^n \to \mathbb{R}$.
- The set of affinely independent vectors of $\mathbb{R}^n$ given by $S_0 = \{x_1, \dots, x_{n+1}\}$ such that $f(x_1) \leq f(x_{i+1})$, and $f(x_{n+1}) \geq f(x_i)$, $i = 1, \dots, n$.
- The required precision $\epsilon \in \mathbb{R}^*_+$.

2: **Initialization:** $k \leftarrow 0$.

3: **Iterations:**

4: Compute S_r, consisting of x_1 and $x_i^r = 2x_1 - x_i$, $i = 2, \dots, n+1$.

5: If $f(S_r) < f(x_1)$:
a) Compute S_e, consisting of x_1 and $x_i^e = 3x_1 - 2x_i = 2x_i^r - x_1$, $i = 2, \dots, n+1$.
b) If $f(S_e) < f(S_r)$, then $S_{k+1} = S_e$.
c) Else $S_{k+1} = S_r$.

6: If $f(S_r) \geq f(x_1)$, compute S_{k+1}, consisting of x_1 and $x_i^c = \frac{x_1 + x_i}{2}$, $i = 2, \dots, n+1$.

7: $k \leftarrow k+1$.

8: Renumber so that x_1 is the best point and x_{n+1} is the worst point in the new simplex.

9: **Stopping criterion:** If $\|x_{n+1} - x_1\| \leq \epsilon$, then $x^* = x_1$.

10: **Output:** An approximation of the solution $x^* \in \mathbb{R}$.

6.12. Application to an identification problem

If we assume that a sheet of metal hardens purely isotropically, one single scalar parameter is needed to describe the evolution of the flow surface. This parameter is called the equivalent plastic deformation, which is calculated as the integral with respect to time of the rate of equivalent plastic deformation:

$$\varepsilon = \int_0^T \dot{\varepsilon} dt. \qquad [6.41]$$

The evolution of the pair $(\sigma_y, \bar{\varepsilon}^p)$ on the surface is described by Swift's hardening law:

$$\sigma_y = k(\varepsilon_0 \bar{\varepsilon}^p)^n,$$

where k is the hardening coefficient, ε_0 is the reference plastic deformation and n is the work hardening coefficient.

These three parameters represent the characteristics of the material. Three other parameters affecting the deformation characteristics are not included in the hardening model: the anisotropy parameters of the material, the coefficient of friction between the tools and the sheet metal and the thickness t of the sheet.

The parameters (k, ε_0, n) are calculated in such a way that the constitutive equations of the plastification surface reproduce the shaping characteristics of the material as accurately as possible. The problem that remains to be solved is to find the best combination of fatigue parameters that minimizes the difference between numerical predictions and experimental results in the following form:

$$\min \| F_{exp} - F_{num} \|_2^2 .$$

This minimization describes the differences between the experimental measurements of the tension and their numerical predicted values on traction test samples. Due to the complexity of the formulas involved, we will develop a numerical minimization strategy based on the Nelder–Mead method [RAD 16].

The technique for identifying the material parameters is based on a coupling between the Nelder–Mead method (Matlab code) and the numerical simulation according to the finite element method via Abaqus/Explicit© [RAD 11]. Other approaches are also possible (see, for example, [SMA 19]). To extract information from the file generated by Abaqus/Explicit©, we use a sophisticated Python script (see Figure 6.2).

Rectangular samples were produced with the following geometric characteristics: thickness = 1.0 mm, width = 12.52 mm and initial length = 100 mm, made of stainless steel (Figure 6.3).

All numerical simulations were performed under controlled displacement conditions with constant velocity $v = 0.1$ mm/s. The predicted force relative to the displacement curves is compared to the experimental results for three orientations in Figure 6.3. At low ductility (step 1), the maximum stress is around 360 MPa, reaching 25% of plastic deformation and the final rupture is obtained at 45% of plastic deformation. At moderate ductility (step 3), the maximum stress is around 394 MPa, reaching 37.2% of plastic deformation, and the final rupture occurs at 53% of plastic deformation.

The best values for the material parameters obtained from the optimization procedure are summarized in Table 6.1. Within these coefficients describing the response (stress vs. plastic deformation), an isotropic nonlinear hardening occurs in such a way that the maximum stress is attained for plastic deformations and the final rupture being obtained at 22% of plastic deformation. The map of plastic deformation in the optimal case is presented in Figure 6.3.

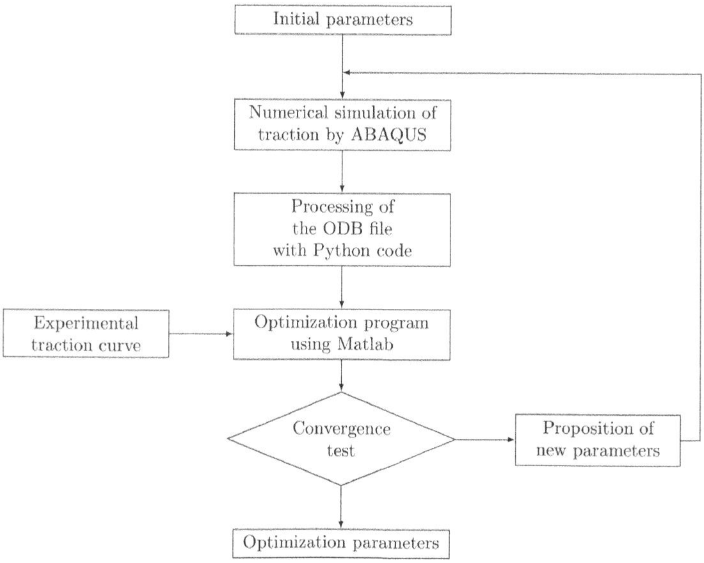

Figure 6.2. *Identification process*

Step	Critical plastic deformation	k [MPa]	ε_0	n
1	25.8%	381.3	0.0100	0.2400
2	29.8%	395.5	0.0120	0.2415
3	37.2%	415.2	0.0150	0.2450
Optimal	36.8%	416.1	0.0198	0.2498

Table 6.1. *Material properties*

6.13. Using Matlab

Optimization toolbox offers various functions with a range of options for solving unconstrained optimization problems, for example:

– *fminunc*: finds the minimum of a function of multiple variables. The solving algorithm can be specified as based on trust regions or Newton-type methods with the following syntax:

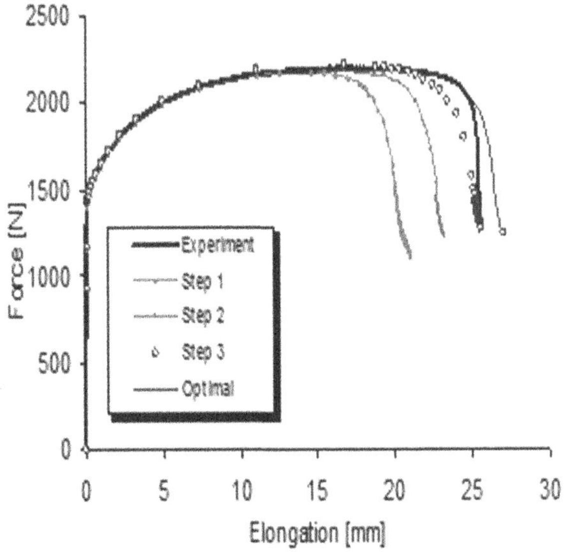

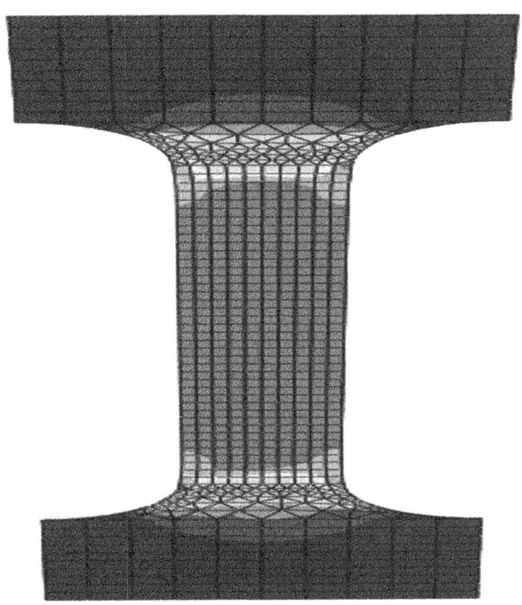

Figure 6.3. *Force/elongation for different optimization steps and plastic deformation plane. For a color version of this figure, see www.iste.co.uk/radi/optimizations.zip*

[x,fval] = fminunc(fun,x0,options) gives the optimal value of the objective function with fval at the optimal point x.

– *fminsearch*: based on the Nelder–Mead algorithm.

6.13.1. *The* fminsearch *function*

Matlab uses the *fminsearch* function (simplex method) because it does not need to compute either the gradient or the Hessian matrix at each iteration. The method works by surrounding the minimum with a simplex. A simplex is a set of $N+1$ points around the minimum (a line in one dimension, a triangle in two dimensions, a pyramid in three dimensions). Each simplex is characterized by $n+1$ vectors at its vertices. At each step of the computation, a new point is selected inside or next to the simplex. The value of the function at this point is compared against the values of the functions evaluated at the vertices, and one of the vertices is typically replaced by the new point, generating a new simplex (by reflection, expansion or contraction). This procedure is repeated until the diameter of the simplex falls below a specified tolerance.

Let us define the function $f(x) = 100(x_2 - x_1^2) + (1 - x_1)^2$ in the file "f.m":

```
function y = f(x)
y=100*(x(2)-x(1)^2)^2+(1-x(1))^2;
```

For a color version of this code, see www.iste.co.uk/radi/optimizations.zip

We can minimize this function as follows in the Matlab command prompt:

```
>> x0 = [5 5];
>> [x, fval, exitflag, output] = fminsearch('f',x0)
```

For a color version of this code, see www.iste.co.uk/radi/optimizations.zip

This gives the following result:

```
x =

    1.0000    1.0000

fval =

   5.6197e-10
```

```
exitflag =

     1
output =
    iterations: 106
     funcCount: 201
     algorithm: 'Nelder-Mead simplex direct search'
       message: 'Optimization terminated:
the current x satisfies the termination criteria using
OPTIONS.TolX of 1.0e-06'
```

6.13.2. *The* fminunc *function*

Another function with a similar syntax is *fminunc*, which solves nonlinear and multivariable optimization problems without restrictions. This function can switch between different algorithms, including the Interior Reflective Newton method (if the derivatives are known) or the BFGS method (if not). The BFGS algorithm approximates the Hessian matrix (quasi-Newton method).

Let us define the function $f(x) = x_1^2 x_2 + x_1 x_2^2 + x_1^2 + x_2^2$ in the file "f.m":

```
function y = f(x)
y= x(1)^2*x(2)+x(1)*x(2)^2+x(1)^2+x(2)^2;
```

For a color version of this code, see www.iste.co.uk/radi/optimizations.zip

For the BFGS algorithm, which approximates the Hessian matrix at each step of the calculation, we enter the following in the Matlab command prompt:

```
>> x0 = [1 -1];
>> [x,fval,exitflag,output,gradient,hessian] =
fminsearch('f',x0)
```

For a color version of this code, see www.iste.co.uk/radi/optimizations.zip

This gives the following result:

```
x =

   -0.0014   -0.3631
fval =
   1.3183e-13
```

```
exitflag =

     1
output =
       iterations: 8
        funcCount: 27
         stepsize: 1
    firstorderopt: 7.4107e-07
        algorithm: 'quasi-newton'
          message: 'Local minimum found.
gradient =

   -0.0178
   -0.7411
hessian =

    2.0000   -0.0002
   -0.0002    2.0000
```

If the gradient is known, we can define it explicitly:

```
function [y,dy] = f(x)
y = x(1)^2*x(2)+x(1)*x(2)^2+x(1)^2+x(2)^2;
dy(1) = 2*x(1)*x(2)+x(2)^2+2*x(1);
dy(2) = x(1)^2+2*x(1)*x(2)+2*x(2);
```

For a color version of this code, see www.iste.co.uk/radi/optimizations.zip

In the Matlab command prompt, we enter:

```
>> options = optimset('GradObj','on');
>> x0 = [1 1];
>> [x,fval,exitflag,output] = fminunc(@f,x0,options)
```

For a color version of this code, see www.iste.co.uk/radi/optimizations.zip

For large problems, Matlab recommends using the "interior reflective Newton" method with preconditioned conjugate gradients, since this algorithm converges more quickly than the other option. For medium-sized problems, the quasi-Newton BFGS method is often used. Matlab also allows us to specify the method of maximum descent, which computes the Hessian matrix: HessUpdate "steepdesc".

6.13.3. *Relaxation method*

Consider the following unconstrained nonlinear minimization problem:

$$\min_{x_1,x_2} f(x_1,x_2) = 2x_1 + x_2 + (x_1^2 - x_2^2) + (x_1 - x_2^2)^2. \quad [6.42]$$

Based on the optimality conditions, the analytic solution is as:

$$x_1^* = -0.185 \quad \text{and} \quad x_2^* = -0.793. \quad [6.43]$$

We can apply the relaxation method with the following script:

```
j = 1;
%
% Initial point
%
x1zero = 3.0;
x2zero = 3.5;
%
% Search intervals
%
a1 = -5.0;
b1 = 5.0;
a2 = -4.0;
b2 = 4.0;
%
% Number of subintervals for local search
%
n = 250.0;
Δ=1.0e3;
epsilon = 1.0e-6;
%
% Open the file "results.txt"
%
fp = fopen('results.txt' , 'w ');
fprintf (fp, ' \n\n Relaxation Method \n ');
fprintf (fp,'==========\n\n ') ;
%
ls1 = a1;
ld1 = b1;
ls2 = a2;
ld2 = b2;
%
fprintf (fp, 'Search limits: \n\n ');
fprintf (fp, '151=%f Idl=%f\n 152=%f Id2=%f\n\n ',ls1, ld1,
            ls2, ld2);
%
% Objective function with the initial point
%
fprintf (fp, 'coordinates of the initial point ' );
fprintf (fp, ' \n\n ' );
fprintf (fp, ' x1zero=%f x2zero=%f\n\n ' ,x1zero, x2zero );
```

```
f_old = f(x1zero,x2zero);
fprintf (fp, ' F (X0) =%f\n\n ' , f_old );
fMin = f_old;
%
% Start the search
%
x2 = x2zero;
%
while(Δ>epsilon)
% Block I: for the variable "x1"
x1 = ls1;
increment = abs(ld1−ls1)/n;
%
while(x1≤ld1)
  func = f ( x1 , x2 );
    if(func<fMin)
     fMin = func;
     x1Min = x1;
  end
  x1 = x1 + increment;
end
%
% Block II: for the variable "x2"
x1 = x1Min;
x2 = ls2;
increment = abs(ls2 −ld2)/n;
%
while(x2≤ld2)
 func = f (x1 , x2 );
  if(func < fMin)
    fMin = func;
    x2Min = x2;
  end
 x2 = x2 + increment;
end
%
f_now - fMin;
fprintf(fp,'i=%d x1=%f x2=%f fMin=%f\n',j, x1Min, x2Min, f_new);
%
% Convergence test
%
Δ = abs(f_new−f_old);
j = j + 1;
f_old = f_new;
x2 = x2Min;
end
%
fprintf(fp, '\n  FINAL RESULT : \n\n');
fprintf(fp, 'x1Min=%f\n x2Min=%f\n\n', x1Min, x2Min);
fprintf(fp, 'fMin=%f\n\n', fMin);
fclose(fp) ;
```

For a color version of this code, see www.iste.co.uk/radi/optimizations.zip

together with the following script that returns the value of the function f:

```
%       f.m
% evaluation of the function f
function [val] = f(x,y)
val = 2*x+y+(x.^2- y.^2 ) + (x-y.^2).^2;
end
%
```

For a color version of this code, see www.iste.co.uk/radi/optimizations.zip

If we run this script with the search intervals $[-5, 5]$ for x_1 and $[-4, 4]$ for x_2 and the initial point $x_0 = (3, 3.5)^T$, the method converges to $x^* = (-0.2, -0.8)^T$ with $z^* = -1.0944$ after eight iterations. A selection of Matlab scripts can be found in [ANC 19].

7

Constrained Nonlinear Optimization

7.1. Introduction

Nonlinear programming (NLP) is primarily concerned with optimization problems whose assumptions (i.e. the functions and sets defining the problem) are nonlinear, but are nevertheless sufficiently differentiable to apply theoretical tools such as optimality conditions, or solving algorithms specifically introduced and analyzed for this purpose. This subfield of optimization is somewhat ambiguously defined around the edges. Its introduction was arguably slightly artificial, and its existence is due to the community of researchers who specialize in these topics and the results that have been successfully established about them [BER 99, BAZ 13, AVR 03].

7.2. Mathematical formulation

Let $U \subset \mathbb{R}^n$, where U is a closed set of constraints, and $U \neq \mathbb{R}^n$, and consider a function $J : \mathbb{R}^n \to \mathbb{R}$.

We wish to solve the minimization problem

$$\min_{u \in U} J(u). \quad [7.1]$$

We will consider two cases for U:

1) U is a block, i.e. of the form $U = [a_1, b_1] \times [a_2, b_2] \times \cdots \times [a_n, b_n]$ (the block can have infinite volume);

2) U is of the form:

$$U = \{x \in \mathbb{R}^n, \phi_i(x) \leq 0,\ i = 1, \in, \ldots, m\}, \quad [7.2]$$

where $m \in \mathbb{N}^*$ and $\phi_1, \phi_2, \ldots, \phi_m$ are known functions from $\mathbb{R}^n$ into $\mathbb{R}$.

REMARK.–

1) The first case is a special case of the second case.

2) In the case of constraints with non-strict inequalities, the corresponding minimization problem is also called a nonlinear programming problem if at least one of the functions $\phi_1, \phi_2, \dots, \phi_m$ or J is not affine. If these functions are all affine, the problem is a linear programming problem.

3) In practice, we might encounter minimization problems with equality constraints, i.e. constraints of the form $\tilde{U} = \{x \in \mathbb{R}^n,\, \phi_i(x) = 0,\, i = 1, 2, \dots, m\}$, or problems with mixed constraints (equalities and non-strict inequalities). If so, we can always reduce to a problem with non-strict inequality constraints.

We have the following result:

THEOREM 7.1.– Let U be given by [7.2]. If the functions $\phi_1, \phi_2, \dots, \phi_m$ are all convex, then U is a convex set.

7.3. Lagrange multipliers

Let $J, \theta_1, \theta_2, \dots, \theta_m \colon \mathbb{R}^n \to \mathbb{R}$ be functions of class C^1, with $m \in \mathbb{N}^*$. Write

$$\tilde{O} = \{x \in \mathbb{R}^n,\, \theta_i(x) = 0,\ i = 1, 2, \dots, m\}.$$

We say that $\tilde{O}$ is a subset of $\mathbb{R}^n$.

DEFINITION 7.1.–

1) If $x \in \tilde{O}$ is such that the family of vectors $\{\nabla\theta_i(x)\}_{i=1,\dots,m}$ forms a free system in $\mathbb{R}^n$, then we say that x is a regular point of $\tilde{O}$.

2) $\tilde{O}$ is said to be regular if all points of $\tilde{O}$ are regular.

We have the following result:

THEOREM 7.2.– Let x^* be a regular point of $\tilde{O}$ such that x^* is a local extremum from J to $\tilde{O}$ (local minimum or local maximum). Then there exist $\lambda_1^*, \lambda_2^*, \dots, \lambda_m^* \in \mathbb{R}$ (called Lagrange multipliers) such that

$$\nabla J(x^*) + \sum_{i=1}^{m} \lambda_i^* \nabla\theta_i(x^*) = 0. \qquad [7.3]$$

The system [7.3] gives necessary conditions of optimality (also known as first-order necessary conditions of optimality). They are not sufficient in general. The system [7.3] gives us n equations with $n + m$ unknowns. But the fact that $x^* \in \tilde{O}$ gives us another m equations: $\theta_i(x^*) = 0$, $i = 1, \dots, m$, which results in $n + m$ equations with $n + m$ unknowns in total.

7.4. Optimization with inequality constraints

Again, consider the functions $J, \theta_1, \theta_2, \ldots, \theta_m \colon \mathbb{R}^n \to \mathbb{R}$ of class C^1, with $m \in \mathbb{N}^*$. Let:

$$O = \{x \in \mathbb{R}^n,\ \theta_i(x) \leq 0,\ \forall i = 1, 2, \ldots, m\}.$$

Notation:

Let us introduce the vector-valued function $\theta \colon \mathbb{R}^n \to \mathbb{R}^m$, given by $\theta(x) = (\theta_1(x), \theta_2(x), \ldots, \theta_m(x))^T$.

For a vector $b = (b_1, b_2, \ldots, b_m)^T \in \mathbb{R}^m$, the notation $b \leq 0$ means that $b_i \leq 0$, $\forall i = 1, \ldots, m$. We can therefore write $O = \{x \in \mathbb{R}^n,\ \theta(x) \leq 0\}$.

Let us consider the minimization problem:

$$\min_{x \in O} J(x). \qquad [7.4]$$

7.4.1. *First-order conditions of optimality*

7.4.1.1. *Lagrange conditions*

In the case of equality constraints, consider the following problem:

$$\begin{cases} \min\ f(x) \\ \text{s.t. } h_j(x) = 0,\ j = 1, \ldots, n \\ \qquad x \in \mathbb{R}^n, \end{cases} \qquad [7.5]$$

where the functions f and h_j are of class C^1. The Lagrange conditions given by the next theorem provide necessary conditions for an element of $\mathbb{R}^n$ to be a local minimum of the problem [7.5]:

THEOREM 7.3.– Let x^* be a local minimum of the problem. Suppose that the constraints are qualified at x^*. Then there exist p real numbers μ_j satisfying $\nabla f(x^*) = \sum_j \mu_j \nabla h_j(x^*)$.

The following result gives hypotheses under which the Lagrange conditions are sufficient.

THEOREM 7.4.– The Lagrange conditions are sufficient whenever f is convex in a neighborhood of x^* and the h_j are affine.

7.4.1.2. *Karush–Kuhn–Tucker conditions*

We will now state optimality conditions similar to the system [7.3], except for inequality constraints.

DEFINITION 7.2.– *We say that the constraint* $\theta_i(u) \leq 0$ *is active at* $v \in O$ *if we have* $\theta_i(v) = 0$. *If so, we introduce the set* $I(v) = \{\theta_i(v) = 0, i = 1, 2, \ldots, m\}$ *and the manifold (only defined if* $I(v)$ *is non-empty)* $\tilde{O}(v) = \{u \in \mathbb{R}^n, \theta_i(u) = 0, \forall i \in I(v)\}$.

DEFINITION 7.3.– *We say that* $v \in O$ *is a regular point of* O *if:*

– *either* $I(v) = \emptyset$;

– *or* v *is a regular point of the manifold* $\tilde{O}(v)$.

We have the following result:

Let x^* be a regular point of O. If x^* is a local minimum point from J to O, then there exist $p_1^*, p_2^*, \ldots, p_m^* \in [0, +\infty[$ (called Karush–Kuhn–Tucker [KKT] multipliers) such that:

$$\nabla J(x^*) + \sum_{i=1}^{m} p_i^* \nabla \theta_i(x^*) = 0 \qquad [7.6]$$

$$p_i^* \theta_i(x^*) = 0, \quad \forall i = 1, 2, \ldots, m. \qquad [7.7]$$

REMARK.– These are once again first-order necessary conditions of optimality, known as KKT conditions.

We have the following result:

THEOREM 7.5.– Let $v \in O$ be such that $I(v)$ is non-empty, and let $w \in \mathbb{R}^n$ be such that $< \nabla\theta_j(v), w >< 0$, for every $j \in I(v)$. Suppose that v is a local minimum from J to O. Then $< \nabla J(v), w >\leq 0$.

DEFINITION 7.4.– *We say that the constraints of* O *are qualified at* $v \in O$ *if*

– *either* $I(v) = \emptyset$;

– *or* $I(v) \neq \emptyset$ *and there exists* $w \in \mathbb{R}^n$ *such that, for* i *in* $I(v)$, *we have* $< \nabla\theta_i(v), w >\leq 0$, *and also* $< \nabla\theta_i(v), w >< 0$, *if* θ_i *is not affine.*

We have the following result:

Let $x^* \in O$ be such that the constraints of O are qualified at x^*. If x^* is a local minimum point from J to O, then there exist $p_1, p_2, \ldots, p_m \in [0, +\infty[$ (called Lagrange multipliers) such that:

$$\nabla J(x^*) + \sum_{i=1}^{m} p_i^* \nabla \theta_i(x^*) = 0 \qquad [7.8]$$

$$p_i^* \theta_i(x^*) = 0, \quad \forall i = 1, \in, \ldots, m. \qquad [7.9]$$

The following result gives a sufficient condition for qualification at every point of O:

THEOREM 7.6.– Suppose that the functions $\theta_1, \theta_2, \ldots, \theta_m$ are convex. Also, suppose that

– either all the functions $\theta_1, \theta_2, \ldots, \theta_m$ are affine;

– or there exists $y \in \mathbb{R}^n$ such that, for every $i = 1, \ldots, m$, we have $\theta_i(y) \leq 0$ or $\theta_i(y) < 0$ if θ_i is not affine.

Then, for every $x \in O$ the constraints of O are qualified at x.

EXAMPLE 7.1.– Let $O_1 = \{x \in \mathbb{R}^2,\ x_2 \geq x_1^2,\ x_2 - x_1 = 4\}$. It is clear that we can write O_1 in the form $O_1 = \{x \in \mathbb{R}^2,\ \theta_i(x) \leq 0,\ i = 1, 2, 3\}$ with $\theta_1, \theta_2, \theta_3 \colon \mathbb{R}^2 \to \mathbb{R}$ defined by

$$\theta_1(x) = x_1^2 - x_2,\ \theta_2(x) = x_2 - x_1 - 4,\ \theta_3(x) = -x_2 + x_1 + 4.$$

Observe that θ_1, θ_2, and θ_3 are convex, and also that θ_2 and θ_3 are affine. Taking $y = (0, 4)^T$, we have $\theta_1(y) < 0$, $\theta_2(y) = \theta_3(y) = 0$. Thus, the hypotheses of the previous theorem are satisfied, so every point of O_1 is qualified.

7.4.2. *Presentation of saddle points*

Let U and P be two arbitrary sets, and consider a mapping $\mathcal{L} \colon U \times P \to \mathbb{R}$.

DEFINITION 7.5.– *We say that the pair $(u^*, p^*) \in U \times P$ is a saddle point of $\mathcal{L}$ if we have*

$$\mathcal{L}(u^*, p) \leq \mathcal{L}(u^*, p^*) \leq \mathcal{L}(u, p^*), \quad \forall u \in U,\ \forall p \in P.$$

EXAMPLE 7.2.– Choose $U = P = \mathbb{R}$ and $\mathcal{L}(u, p) = u^2 - p^2$. It is easy to see that $(0, 0)$ is a saddle point of $\mathcal{L}$.

We introduce the following functions: $G\colon U \to \overline{\mathbb{R}}$ and $H\colon P \to \overline{\mathbb{R}}$ given by $G(u) = \sup_{q\in P} \mathcal{L}(u,q)$ and $H(p) = \inf_{v\in U} \mathcal{L}(v,p)$, respectively.

The point (u^*, p^*) is a saddle point of $\mathcal{L}$ if and only if $G(u^*) = \mathcal{L}(u^*, p^*) = H(p^*)$.

We have the following result:

THEOREM 7.7.– $\sup\limits_{p\in P} \inf\limits_{u\in U} \mathcal{L}(u,p) \leq \inf\limits_{u\in U} \sup\limits_{p\in P} \mathcal{L}(u,p)$.

If (u^*, p^*) is a saddle point from $\mathcal{L}$ to $U \times P$, then

$$\sup_{p\in P} \inf_{u\in U} \mathcal{L}(u,p) = \mathcal{L}(u^*, p^*) = \inf_{u\in U} \sup_{p\in P} \mathcal{L}(u,p). \qquad [7.10]$$

EXAMPLE 7.3.– Again, consider the previous example: $\mathcal{L}\colon \mathbb{R}\times\mathbb{R} \to \mathbb{R}$, $\mathcal{L}(u,p) = u^2 - p^2$. We have $G(u) = \sup\limits_{p\in\mathbb{R}}(u^2 - p^2) = u^2$, and $H(p) = \inf\limits_{u\in\mathbb{R}}(u^2 - p^2) = -p^2$. We also have $\inf\limits_{u\in\mathbb{R}} G(u) = 0$, $\sup\limits_{p\in\mathbb{R}} H(p) = 0$ and $\mathcal{L}(0,0) = 0$.

7.4.3. *Saddle point and optimization*

We introduce the Lagrangian $\mathcal{L}\colon \mathbb{R}^n \times \mathbb{R}^m_+ \to \mathbb{R}$, given by:

$$\mathcal{L}(x,p) = J(x) + \sum_{i=1}^{m} p_i\theta_i(x) \equiv J(x) + < p, \theta(x) > . \qquad [7.11]$$

We have the following result:

THEOREM 7.8.– If (x^*, p^*) is a saddle point from $\mathcal{L}$ to $\mathbb{R}^n \times \mathbb{R}^m_+$, then x^* is a minimum from J to O. Furthermore, (x^*, p^*) satisfies the KKT conditions, namely

$$\begin{cases} \nabla J(x^*) + \displaystyle\sum_{i=1}^{m} p_i^* \nabla\theta_i(x^*) = 0 \\ p_i^*\theta_i(x^*) = 0, \quad \forall i = 1, \ldots, m. \end{cases} \qquad [7.12]$$

This result gives a sufficient condition for a minimum from J to O. We will also see a converse result with additional hypotheses. The converse asks to solve the problem known as the dual problem:

$$\sup_{p\in\mathbb{R}^m_+} \inf_{u\in\mathbb{R}^n} \mathcal{L}(u,p),$$

which is simpler to solve than the problem of minimizing from J to O (known as the primal problem).

EXAMPLE 7.4.– Consider a homogeneous beam with a circular cross-section of diameter $2a$. The beam is fixed on one end and free on the other; there is a load on the free end. Suppose that we wish to find the dimensions of a beam with a rectangular cross-section (obtained by cutting the circular cross-section) that will support the maximum load at the free end.

Let M be the bending moment, I the moment of inertia and σ the stress at distance y. We have

$$\frac{\sigma}{y} = \frac{M}{I},$$

so

$$\sigma_{max} = \frac{M}{I}y = \frac{3M}{4xy^2}.$$

The problem reduces to minimizing the quantity $\frac{k}{xy^2}$ subject to the constraint $x^2 + y^2 = a^2$, which we can express as follows:

$$\begin{cases} \min \quad f(x,y) = kx^{-1}y^{-2} \\ \text{s.t.} \quad g(x,y) = x^2 + y^2 - a^2 = 0 \end{cases} \qquad [7.13]$$

Computing the various gradients gives

$$\frac{\partial f}{\partial x} = -kx^{-2}y^{-2} \ \frac{\partial f}{\partial y} = -kx^{-1}y^{-3},$$

$$\frac{\partial g}{\partial x} = 2x, \qquad \frac{\partial g}{\partial y} = 2y.$$

Thus, we find

$$\left(\frac{\partial f}{\partial x}.\frac{\partial g}{\partial y} - \frac{\partial f}{\partial y}.\frac{\partial g}{\partial x}\right)_{x^*,y^*} = 0,$$

so

$$-2x^{-2}y^{-2}(2y) + 2kx^{-1}y^{-3}(2x) = 0.$$

Hence, we have $y^* = \sqrt{2}x^*$.

Let us solve the problem [7.13]. The Lagrangian is as follows:

$$\begin{aligned} L(x,y,\lambda) &= f(x,y) + \lambda g(x,y) \\ &= kx^{-1}y^{-2} + \lambda(x^2 + y^2 - a^2). \end{aligned}$$

The various partial derivatives of the Lagrangian are:

$$\frac{\partial L}{\partial x} = -kx^{-2}y^{-2} + 2x\lambda = 0$$
$$\frac{\partial L}{\partial y} = -2kx^{-1}y^{-3} + 2y\lambda = 0$$
$$\frac{\partial L}{\partial \lambda} = x^2 + y^2 - a^2 = 0$$

so

$$2\lambda = \frac{k}{x^3y^2} = \frac{2k}{xy^4},$$

and

$$y^* = \sqrt{2}x^*.$$

Hence, we find the same optimum.

EXAMPLE 7.5.– Suppose that we wish to find the dimensions of a cylindrical steel container that maximizes its volume assuming a surface area of $S_0 = 24\pi$. Let x_1 be the radius of the base, and x_2 the thickness. The problem can be expressed as follows:

$$\begin{cases} \max\ f(x_1, x_2) = \pi x_1^2 x_2 \\ \text{s.t.}\ \ g(x_1, x_2) = 2\pi x_1^2 + 2\pi x_1 x_2 = S_0 = 24\pi \end{cases}$$

The Lagrangian is

$$L(x_1, x_2, \lambda) = \pi x_1^2 x_2 + \lambda(2\pi x_1^2 + 2\pi x_1 x_2 - S_0).$$

The optimality conditions are:

$$\frac{\partial L}{\partial x_1} = 2\pi x_1 x_2 + 4\pi\lambda x_1 + 2\pi\lambda x_2 = 0$$
$$\frac{\partial L}{\partial x_2} = \pi x_1^2 + 2\pi\lambda x_1 = 0$$
$$\frac{\partial L}{\partial \lambda} = 2\pi x_1^2 + 2\pi x_1 x_2 - S_0 = 0$$

Hence,

$$\lambda = -\frac{x_1 x_2}{2x_1 + x_2} = -\frac{1}{2}x_1,$$

which implies $x_1 = \frac{x_2}{2}$. Therefore, we have

$$x_1^* = \sqrt{\frac{S_0}{6\pi}};\ x_2^* = \sqrt{\frac{2S_0}{3\pi}} \text{ and } \lambda = -\sqrt{\frac{S_0}{24\pi}},$$

and the maximum for f is $f^* = \sqrt{\frac{S_0^3}{54\pi}}$. If $S_0 = 24\pi$, then

$$x_1^* = 2,\ x_2^* = 4,\ \lambda = -1 \text{ and } f^* = 16\pi.$$

7.4.4. *Convex case*

In this section, we will see converses of the two last theorems in the case where J and θ_j are convex.

THEOREM 7.9.– Suppose that the functions $J, \theta_1, \theta_\in, \ldots, \theta_m$ are all convex. Also, suppose that there exist $p_1^*, p_2^*, \ldots, p_m^* \in [0, +\infty[$ and $x^* \in O$ such that the KKT conditions [7.14] are satisfied. Then

1) x^* is a minimum point from J to O;

2) (x^*, p^*) is a saddle point of $\mathcal{L}$.

The next result gives a connection between saddle points and the KKT conditions. Suppose that the functions $J, \theta_1, \ldots, \theta_m$ are convex. Then (x^*, p^*) is a saddle point of $\mathcal{L}$ if and only if $x^* \in O$ and the conditions [7.14] are satisfied.

The above theorem has a converse:

THEOREM 7.10.– Suppose that the functions J, $\theta_1, \ldots, \theta_m$ are convex. Let x^* be a minimum point from J to O such that the constraints of O are qualified at x^*. Then, there exists $p^* \in \mathbb{R}_+^m$ such that (x^*, p^*) is a saddle point of $\mathcal{L}$.

7.5. Constrained minimization algorithms

The goal of this section is to state a selection of numerical algorithms for solving constrained minimization problems:

$$\min_{u \in U} J(u), \qquad [7.14]$$

where $U \subset \mathbb{R}^n$ is a constraint set.

Recall that this chapter considers two types of constraint:

1) U is a block;

2) U is of the form $U = \{x \in \mathbb{R}^n,\ \theta_i(x) \leq 0, i = 1, 2, \ldots, m\}$, where $m \in \mathbb{N}^*$ and $\theta_1, \theta_2, \ldots \theta_m$ are known functions from $\mathbb{R}^n$ into $\mathbb{R}$.

In the first case, we will use natural extensions of the minimization methods on $\mathbb{R}^n$ presented in Chapter 6. In the second case, we will use duality-based methods.

7.5.1. *Relaxation method*

Suppose that

$$U = [a_1, b_1] \times [a_2, b_2] \times \ldots \times [a_n, b_n] \equiv \prod_{i=1}^{n} [a_i, b_i]. \qquad [7.15]$$

In this case, the algorithm is similar to the relaxation algorithm for unconstrained minimization.

Algorithm 7.1 Relaxation algorithm

Let $u^{(k)} = (u_1^{(k)}, u_2^{(k)}, \ldots, u_n^{(k)})^T$ be given

Compute $u^{(k+1)} = (u_1^{(k+1)}, u_2^{(k+1)}, \ldots, u_n^{(k+1)})^T$ in n successive steps:

$$J(u_1^{(k+1)}, u_2^{(k)}, \ldots, u_n^{(k)}) = \min_{y \in [a_1, b_1]} J(y, u_2^{(k)}, \ldots u_n^{(k)})$$

$$J(u_1^{(k+1)}, u_2^{(k+1)}, u_3^{(k)}, \ldots, u_n^{(k)}) = \min_{y \in [a_2, b_2]} J(u_1^{(k+1)}, y, u_3^{(k)}, \ldots, u_n^{(k)})$$

$$\vdots$$

$$J(u_1^{(k+1)}, \ldots, u_n^{(k+1)}) = \min_{y \in [a_n, b_n]} J(u_1^{(k+1)}, \ldots, u_{n-1}^{(k+1)}, y)$$

Like in the unconstrained case, we have the following convergence theorem:

THEOREM 7.11.– If $J\colon \mathbb{R}^n \to \mathbb{R}$ is an elliptic function and the set U is given by [7.15], then the relaxation method for minimizing from J to U is well defined and converges.

7.5.2. *Projection method*

Recall the results of the projection theorem onto a closed convex set: let $U \subset \mathbb{R}^n$ be a closed, convex, non-empty set and let $v \in \mathbb{R}^n$. Then there exists a unique element $P_U(v) \in U$ such that

$$\|v - P_U(v)\| = \inf_{u \in U} \|v - u\| = \min_{u \in U} \|v - u\|.$$

The element $P_U(v) \in U$ is called the projection of v onto U in $\mathbb{R}^n$. Furthermore:

1) The element $P_U(v) \in U$ also satisfies

$$< v - P_U(v), u - P_U(v) > \leq 0, \forall u \in U. \qquad [7.16]$$

2) If $w \in U$ satisfies

$$< v - w, u - w > \leq 0, \forall u \in U, \qquad [7.17]$$

then $w = P_U(v)$ (so $w = P_U(v)$ is the unique element of U satisfying [7.17]).

3) We have

$$\|P_U(v_1) - P_U(v_2)\| \leq \|v_1 - v\|, \quad \forall v_1, v_2 \in \mathbb{R}^n. \qquad [7.18]$$

The function P_U is Lipschitz and therefore continuous.

4) We have $v = P_U(v)$ if and only if $v \in U$.

5) If U is the closed affine subspace of $\mathbb{R}^n$ given by $U = a + U_0$ with $a \in \mathbb{R}^n$, and U_0 is a closed vector subspace of $\mathbb{R}^n$, then the inequality [7.16] becomes the equality

$$< v - P_U(v), u >= 0, \quad \forall u \in U_0. \qquad [7.19]$$

In the case where U is given by general inequality constraints, it may be difficult in practice to calculate the projection onto U in $\mathbb{R}^n$.

Let us return to the minimization problem [7.14]. In the following, we assume that the constraint set U is convex and closed. Suppose that $x^* \in U$ is such that $J(x^*) = \min_{x \in U} J(x)$. Then

$$< \nabla J(x^*), x - x^* > \geq 0, \ \forall x \in U, \qquad [7.20]$$

which is equivalent to the following inequality for every $\rho > 0$:

$$\rho < -\nabla J(x^*), x - x^* > \leq 0, \quad \forall x \in U. \qquad [7.21]$$

This gives

$$< [x^* - \rho \nabla J(x^*)] - x^*, x - x^* > \leq 0, \quad \forall x \in U. \qquad [7.22]$$

By the projection theorem, this inequality is equivalent to the equality $x^* = P_U(x^* - \rho \nabla J(x^*))$. Therefore, the solution of the minimization problem [7.14] is a fixed point of the mapping $g \colon U \to O$ defined by

$$g(x) = P_U(x - \rho \nabla J(x)).$$

A natural idea for approximating this fixed point numerically, is to use an iterative method that constructs a sequence $x^{(k)} \in U$ recursively:

$$x^{(k+1)} = P_U(x^{(k)} - \rho \nabla J(x^{(k)})), \quad k \in \mathbb{N}, \qquad [7.23]$$

for some fixed $\rho > 0$. We can alternatively vary ρ at each step. Accordingly, the general algorithm will be as follows for this case:

Algorithm 7.2 Gradient method with variable step projection

$$\begin{cases} x^{(k+1)} = P_U(x^{(k)} - \rho_k \nabla J(x^{(k)})), & k \in \mathbb{N} \\ x^{(0)} \text{ given } \in U \end{cases} \quad [7.24]$$

for some given $\rho_k > 0$.

This is the gradient projection method with variable step size. It is called the gradient projection method with fixed step size if ρ_k is independent of k.

Let $U \subset \mathbb{R}^n$ be a closed, convex, non-empty set, and let $J \colon \mathbb{R}^n \to \mathbb{R}$ be an elliptic function such that ∇J is Lipschitz. Thus, there exist $M > 0$ and $\alpha > 0$ such that $\forall x, y \in \mathbb{R}^n$,

$$< \nabla J(x) - \nabla J(y), x - y > \geq \alpha \|x - y\|^2$$

and

$$\|\nabla J(x) - \nabla J(y)\| \leq M\|x - y\|.$$

Suppose that there exist real numbers a_1 and a_2 satisfying

$$0 < a_1 \leq a_2 < \frac{2\alpha}{M^2}$$

such that $a_1 \leq \rho_k \leq a_2$ $\forall k \in \mathbb{N}$. Then the method [7.2] converges geometrically: there exist $C \geq 0$ and $\beta \in [0, 1[$ such that $\|x^{(k)} - x^*\| \leq C\beta^k, \forall k \in \mathbb{N}$.

This method is only applicable if we can easily calculate the projection P_U, for example if U is a block in $\mathbb{R}^n$. For a set U defined by non-strict inequality constraints, it is not easy to apply this method in general.

7.5.3. *Exterior penalty method*

Consider again the minimization problem [7.14]. We introduce a function $\psi \colon \mathbb{R}^n \to \mathbb{R}$ with the following properties:

$$\begin{cases} \psi & \text{continuous and convex function} \\ \psi(x) \geq 0 & \forall x \in \mathbb{R}^n \\ \psi(x) = 0 & \text{if and only if } x \in U \end{cases} \quad [7.25]$$

A function with the properties of [7.25] is called an exterior penalty function for U. For every $\epsilon > 0$, we now introduce a function $J_\epsilon \colon \mathbb{R}^n \to \mathbb{R}$ defined by

$$J_\epsilon(x) = J(x) + \frac{1}{\epsilon}\psi(x). \quad [7.26]$$

The function J_ϵ is called the penalized function of J.

Applying the exterior penalty method to the problem [7.14] means minimizing the function J_ϵ over $\mathbb{R}^n$ with an $\epsilon > 0$ that tends to 0. The term $\frac{1}{\epsilon}\psi(x)$ penalizes the separation of x from U.

This reduces the constrained minimization problem (from J to U) to an unconstrained minimization problem (from J_ϵ to $\mathbb{R}^n$). We can now apply any algorithm from Chapter 6 to solve the unconstrained minimization problem from J_ϵ.

We have the following convergence result:

THEOREM 7.12.– Let $J\colon \mathbb{R}^n \to \mathbb{R}$ be a continuous and strictly convex function that is coercive on $\mathbb{R}^n$, $U \subset \mathbb{R}^n$ a closed, convex, non-empty set, and $\psi\colon \mathbb{R}^n \to \mathbb{R}$ an exterior penalty function for U. Then, for every $\epsilon > 0$, there exist a unique element $x_\epsilon \in \mathbb{R}^n$ satisfying

$$J_\epsilon(x_\epsilon) = \min_{x \in \mathbb{R}^n} J_\epsilon(x).$$

Furthermore, we have $x_\epsilon \to x^*$ for $\epsilon \to 0$, where x^* is the unique point of U satisfying $J(x^*) \leq J(x)$ for every x in U.

REMARK.– This method is useful when it is easy to construct a function with the properties in [7.25].

EXAMPLE 7.6.– The distance of a point $x \in \mathbb{R}^n$ from the (closed and convex) set U, denoted dist(x, U), is defined as the quantity

$$\text{dist}(x, U) = \|x - P_U(x)\|.$$

We then define the function $\psi\colon \mathbb{R}^n \to \mathbb{R}$ by

$$\psi(x) = \text{dist}^2(x, U) = \|x - P_U(x)\|^2.$$

The properties in [7.25] are all easy to show, except the convexity of ψ.

7.5.4. *Uzawa's algorithm*

Uzawa's algorithm is based on the dual formulation and searching for saddle points. It works well for constraint sets defined by non-strict inequalities. Here, we will consider constraint sets of the form $U = O = \{x \in \mathbb{R}^n,\ \theta_i(x) \leq 0,\ \forall i = 1, 2, \ldots, m\}$ with $\theta_1, \theta_2, \ldots, \theta_m\colon \mathbb{R}^n \to \mathbb{R}$, $m \in \mathbb{N}^*$. We can also write $U = O = \{x \in \mathbb{R}^n,\ \theta(x) \leq 0\}$.

In this part, we assume the following hypotheses:

1) J is elliptic, so J is of class $\mathcal{C}^1$ and there exist $\alpha > 0$ such that

$$< J(x) - J(y), x - y > \geq \alpha \parallel x - y \parallel^2, \ \forall x, y \in \mathbb{R}^n.$$

2) For every $k = 1, \ldots, m$, we have θ_k as a convex function of class $\mathcal{C}^1$. Furthermore, θ is Lipschitz, so there exists $M > 0$ such that

$$\parallel \theta(x) - \theta(y) \parallel \leq M \parallel x - y \parallel, \ \forall x, y \in \mathbb{R}^n.$$

3) The constraints of O are qualified at every point $x \in O$.

We introduce the mapping $\mathcal{L}$: $\mathbb{R}^n \times \mathbb{R}^m_+ \to \mathbb{R}$ given by

$$\mathcal{L}(x, p) = J(x) + < p, \theta(x) >= J(x) + \sum_{k=1}^{m} p_k \theta_k(x), \ \forall x \in \mathbb{R}^n, \ \forall p \in \mathbb{R}^m_+. \tag{7.27}$$

Since the set U is closed and convex and J is elliptic, there exists a unique minimum point $x^* \in U$ from J to U. Therefore, there exists $p^* = (p_1^*, p_2^*, \ldots, p_m^*)^T \in \mathbb{R}^m_+$ such that (x^*, p^*) is a saddle point of $\mathcal{L}$. We deduce that

$$\sup_{p \in \mathbb{R}^m_+} \inf_{x \in \mathbb{R}^n} \mathcal{L}(x, p) = L(x^*, p^*) = \inf_{x \in \mathbb{R}^n} \sup_{p \in \mathbb{R}^m_+} \mathcal{L}(x, p). \tag{7.28}$$

Uzawa's algorithm is given as Algorithm 7.3 on the next page.

Uzawa's algorithm has the following convergence result:

THEOREM 7.13.– Let $a_1, a_2 \in \mathbb{R}$ be such that $0 < a_1 \leq a_2 < \frac{2\alpha}{M^2}$. Suppose that the factors ρ_k are such that $a_1 \leq \rho_k \leq a_2$, for every natural integer k. Then the sequence $x^{(k)}$ generated by Uzawa's algorithm converges to the unique solution x^* of the minimization problem [7.14].

7.6. Newton algorithms: SQP method

Newton algorithms are based on linearizing certain equations that characterize the desired solutions, which are deduced from the first-order optimality conditions. These algorithms are primal-dual in the sense that they generate both a primal sequence (x_k) that converges to a solution $\bar{x}$ of the problem and a dual sequence (λ_k) of multipliers that converges to an optimal multiplier $\bar{\lambda}$ associated with $\bar{x}$.

Algorithm 7.3 Uzawa's algorithm

Step 1. Set $k = 0$, choose $p^{(0)} \in \mathbb{R}_+^m$ (for example, $p^{(0)} = 0$) and choose $k_{max} \in \mathbb{N}$ sufficiently large (for example, $k_{max} = 1,000$) and $\epsilon > 0$ sufficiently small (for example, $\epsilon = 10^{-6}$).

Step 2. Compute $x^{(k)} \in \mathbb{R}^n$ that minimizes the function $x \in \mathbb{R}^n \to J(x) + \sum_{j=1}^m p_j^{(k)} \theta_j(x)$ over $\mathbb{R}^n$ (apply one of the unconstrained minimization algorithms from Chapter 6).

If ($k \geq 1$), then

If $\| x^{(k)} - x^{(k-1)} \| \leq \epsilon$, then $x^{(k)}$ is the desired minimum point (the algorithm has converged)

Else go to step 3.

Step 3.

If ($k = k_{max}$), then the algorithm diverges.

Else do:

$$p_i^{(k+1)} = \max\{p_i^{(k)} + \rho_k \theta_i(x^{(k)}), 0\} \quad i = 1, \ldots, m$$

then $k = k + 1$, return to step 2.

7.6.1. *Equality constraints*

Consider a (differentiable) optimization problem with equality constraints:

$$\begin{cases} \min\limits_{x \in \mathbb{R}^n} f(x) \\ \text{s.t.} \quad h_i(x) = 0, \ i = 1, \ldots, p. \end{cases} \qquad [7.29]$$

where $f \colon \mathbb{R}^n \to \mathbb{R}$ and $h_i \colon \mathbb{R}^n \to \mathbb{R}$, $i = 1, \ldots, p$, are assumed to be at least differentiable.

The Lagrange optimality conditions are:

$$\nabla \mathcal{L}(x, \lambda) = 0 \quad \text{i.e.:} \quad \begin{pmatrix} \nabla f(x) + \sum_{i=1}^p \lambda_i \nabla h_i(x) \\ h(x) \end{pmatrix} = \begin{pmatrix} 0 \\ 0 \end{pmatrix}.$$

Let us use Newton's method to solve this system of equations. One of its iterations is as follows:

$$H[L](x^k, \lambda^k) \begin{pmatrix} x^{k+1} - x^k \\ \lambda^{k+1} - \lambda^k \end{pmatrix} = -\nabla \mathcal{L}(x^k, \lambda^k), \qquad [7.30]$$

which is:

$$\begin{pmatrix} H_x[L](x^k, \lambda^k) & Dh(x^k)^T \\ Dh(x^k) & 0 \end{pmatrix} \begin{pmatrix} x^{k+1} - x^k \\ \lambda^{k+1} - \lambda^k = - \begin{pmatrix} \nabla_x L(x^k, \lambda^k) \\ h(x^k) \end{pmatrix} \end{pmatrix}, \qquad [7.31]$$

where $Dh(x)$ denotes the Jacobian matrix of the mapping $h\colon \mathbb{R}^n \to \mathbb{R}^p$, defined by $Dh(x)^T = [\nabla h_1(x)| \dots |\nabla h_p(x)]$.

Set $H_k = H_x[L](x^k, \lambda^k)$, $d = x^{k+1} - x^k$ and $\mu = \lambda^{k+1}$. The iteration [7.30] therefore becomes:

$$\begin{pmatrix} H_k & Dh(x^k)^T \\ Dh(x^k) & 0 \end{pmatrix} \begin{pmatrix} d \\ \mu - \lambda^k = - \begin{pmatrix} \nabla_x L(x^k, \lambda^k) \\ h(x^k) \end{pmatrix} \end{pmatrix} \qquad [7.32]$$

and is well defined provided that the matrix $H[L](x^k, \lambda^k)$ is invertible. This is indeed the case if:

1) the columns $\nabla h_1(x), \dots, \nabla h_p(x)$ of $Dh(x^k)^T$ are linearly independent: this is the constraint qualification hypothesis;

2) for any $d \neq 0$ such that $Dh(x^k)d = 0$, we have $d^T H_k d > 0$: this is the second-order sufficient condition of optimality for equality constraints.

Let us return to the iteration [7.30]. It can also be written as follows:

$$\begin{cases} H_k d + \sum_{i=1}^{p} (\mu_i - \lambda_i^k) \nabla h_i(x^k) = -\nabla_x \mathcal{L}(x^k, \lambda^k) \\ \nabla h_i(x^k)^T d + h_i(x^k) = 0 & i = 1, \dots, p \end{cases}$$

But $\nabla_x \mathcal{L}(x^k, \lambda^k) = \nabla f(x^k) + \sum_{i=1}^{p} \lambda_i^k \nabla h_i(x^k)$, so:

$$\begin{cases} H_k d + \sum_{i=1}^{p} \mu_i \nabla h_i(x^k) = -\nabla f(x^k) \\ \nabla h_i(x^k)^T d + h_i(x^k) = 0 & i = 1, \dots, p \end{cases}$$

This system can be recognized as the Lagrange optimality conditions of the following quadratic problem:

$$(QP_k) \begin{cases} \min_{d \in \mathbb{R}^n} & \nabla f(x^k)^T d + \frac{1}{2} d^T H_k d \\ \text{s.t.} & h_i(x^k) + \nabla h_i(x^k)^T d = 0. \end{cases} \qquad [7.33]$$

The problem (QP_k) can be seen as the problem of minimizing a quadratic approximation of the Lagrangian of PE with a linear approximation of the constraints.

The Sequential Quadratic Programming (SQP) method proceeds by replacing the initial problem with a sequence of quadratic problems with linear constraints that are easier to solve (see Algorithm 7.4).

Algorithm 7.4 SQL algorithm with equality constraints

1: **Assumptions**: $f\colon \mathbb{R}^n \to \mathbb{R}$, $h\colon \mathbb{R}^n \to \mathbb{R}^p$ differentiable, x_0 initial point, $\lambda^0 \in \mathbb{R}^p$ initial multiplier, $\epsilon > 0$ required precision.
2: $k \leftarrow 0$.
3: **While** $\| \nabla\mathcal{L}(x^k, \lambda^k) \| > \epsilon$,

1) Solve the quadratic subproblem:

$$(QP_k) \left\{ \begin{array}{ll} \min\limits_{d\in\mathbb{R}^n} & \nabla f(x^k)^T d + \frac{1}{2} d^T H_k d \\ \text{s.t.} & h_i(x^k) + \nabla h_i(x^k)^T d = 0. \end{array} \right.$$

and obtain the primal solution d_k and the multiplier λ' associated with the equality constraint.

2) $x^{k+1} = x^k + d_k$; $\lambda^{k+1} = \lambda'$.

4: $k \leftarrow k + 1$.
5: Return x^k.
6: **Output**: an approximation x^* of the solution.

7.6.2. *Inequality constraints*

We will now consider problems with both equality and inequality constraints:

$$\left\{ \begin{array}{ll} \min\limits_{x\in\mathbb{R}^n} & f(x) \\ \text{s.t.} & h_i(x) = 0, \; i = 1, \ldots, p, \\ & g_j(x) \leq 0, \; j = 1, \ldots, q. \end{array} \right. \qquad [7.34]$$

As we did for a problem with only equality constraints, we can linearize the constraints and use a quadratic approximation of the Lagrangian:

$$\mathcal{L}(x, \lambda, \mu) = f(x) + \lambda^T g(x) + \mu^T h(x), \text{ with } \lambda_j \geq 0, \; j = 1, \ldots, q. \qquad [7.35]$$

Algorithm 7.5 SQP algorithm with equality and inequality constraints

1: **Assumptions:** $f\colon \mathbb{R}^n \to \mathbb{R}$, $g\colon \mathbb{R}^n \to \mathbb{R}^q$, $h\colon \mathbb{R}^n \to \mathbb{R}^p$ differentiable, x_0 initial point, $\lambda_0 \in \mathbb{R}^q_+$ and $\mu_0 \in \mathbb{R}^p$ initial multipliers, $\epsilon > 0$ required precision.
2: $k \leftarrow 0$.
3: **While** $\| \nabla \mathcal{L}(x^k, \lambda^k) \| > \epsilon$,

1) Solve the quadratic subproblem:

$$(QP_k) \begin{cases} \min_{d \in \mathbb{R}^n} & \nabla f(x^k)^T d + \frac{1}{2} d^T H_k d \\ \text{s.t.} & h_i(x^k) + \nabla h_i(x^k)^T d = 0,\ i = 1, \ldots, p, \\ & g_j(x^k) + \nabla g_j(x^k)^T d = 0,\ j = 1, \ldots, q. \end{cases}$$

and obtain the primal solution d_k and the multipliers λ' and μ' associated with the inequality and equality constraints, respectively.

2) $x^{k+1} = x^k + d_k$; $\lambda^{k+1} = \lambda'$; $\mu^{k+1} = \mu'$.

4: $k \leftarrow k + 1$.
5: **Return** x^k.
6: **Output:** an approximation x^* of the solution.

REMARK.– To ensure that the quadratic subprogram (QP) admits a unique solution, the most current implementations of SQP use an approximation of the Hessian H_k of the Lagrangian that is positive definite, such as the one provided by quasi-Newton techniques (BFGS).

As a Newton method, the SQP algorithm exhibits quadratic convergence locally, provided that the initial points (x_0, λ_0) (respectively, (x_0, λ_0, μ_0)) are in the neighborhood of a stationary point $\bar{x}$ and of the associated multipliers $\bar{\lambda}$ (respectively, $(\bar{\lambda}, \bar{\mu})$).

7.7. Application to structure optimization

Structure optimization is a key concern when designing systems in mechanical engineering and an extremely important problem for engineers. Not only do engineering departments attempt to improve the mechanical performance of their parts in terms of resistance, lifetime and usage properties, they also optimize the weight, volume and production cost of these parts. This challenge is encountered every day in cutting-edge industries, whether in space research, aeronautics, automotives, naval competitions, fine mechanics (precision mechanics) or civil engineering artworks [ELA 09, ELH 11, ELH 13].

Initiated by the large manufacturers in the transportation sector, structure optimization is now widely used by various industries, such as tooling, aerospace, automotive, and so on. Lying at the intersection of design, aesthetics and ergonomics, this field has become an industrial priority for research and development. As a result, new computational programs capable of optimizing the shape of a structure have emerged among the software used by the mechanical engineering industry [ELH 13].

Mathematically, a structure optimization problem is defined by three assumptions:

– a model, typically a partial differential equation that allows us to evaluate the mechanical behavior of a structure;

– a criterion to be maximized or minimized, or possibly several criteria;

– an admissible set of optimization variables that describes any constraints imposed on the variables.

Structure optimization problems can be classified into three major categories according to the nature of the design variables:

– Sizing optimization: The variables are sections, thicknesses and so on; any quantity that can evolve without requiring a modification of the finite elements mesh. The shape is fixed.

In mathematics, the goal of design optimization is to find the design parameters that minimize an objective function. The general formulation of design optimization can be stated as follows:

$$\begin{array}{l} \min_d \; G_I(d) \\ \text{s.t.} \quad \begin{cases} G_i(d, X) \geq 0 & i = 1, \ldots, m \\ h_j \geq 0 & j = m+1, \ldots, n_h \end{cases} \end{array} \qquad [7.36]$$

where d is the vector of variables being optimized, called the decision variables; X is the vector of deterministic parameters; G_i are the performance functions; h_j are the geometric and physical feasibility functions; and G_I is the objective function.

– Shape optimization: Changes are permitted to the shape if they are compatible with a topology chosen beforehand [ELH 10]. The general approach of a shape optimization problem is as follows:

- find the existence of an optimal solution, and possibly show the uniqueness of this optimal solution (although this is a very difficult question in general);

- study the optimality conditions that characterize optimal shapes;

- finally, compute the optimal solution, or at least satisfactory approximations.

– Topology optimization: The nature of the structure is allowed to vary more fundamentally. Mathematically, the topology optimization problem is of the form [MAK 08]:

$$\min_{\omega \subset \Omega} f(\omega) \qquad [7.37]$$

subject to the constraints:

$$g_i(\omega) \leq 0 \quad 1 \leq i \leq m$$
$$h_j(\omega) = 0 \quad 1 \leq j \leq n$$

where f is the objective function and the g_i and the h_j are functions defining the constraints. In practice, these functions may be implicit and nonlinear in ω, which means that a state equation needs to be solved to evaluate them.

In most optimization problems encountered by engineers, a systematic approach is required, consisting of the following four key phases.

– *Phase I*: Precise definition of the specifications

This phase establishes a rigorous characterization of the problem being optimized. The following aspects are defined precisely:

1) the desired behavior and performance (objectives);

2) the design parameters that will influence the performance (optimization variables, also called design variables);

3) any constraints associated with the problem.

This is a key phase of the optimization process with extensive influence on subsequent stages. It is not easy to approach, as the choice of design variables is never unique. In general, it is advisable to define a problem as simply as possible with the fewest number of parameters. Having a well-posed problem with sensibly chosen parameters makes solving it much easier.

– *Phase II*: Formulation in terms of mathematical equations

This phase involves translating the physical problem into an equivalent mathematical problem. It is undoubtedly the most delicate step of the optimization process, since the mathematical formulation of a problem is, once again, never unique. Starting from a physical problem, the precise setting of the study is defined with simplifying hypotheses that allow a suitable mathematical model to be determined. To achieve this objective, we need to choose which laws of physics will yield equations that translate the reality of the problem with the desired level of precision.

If the mathematical model does not have an analytic solution, an approximate solution of the model is required. Discretizing the problem then corresponds to choosing a numerical model that will allow the mathematical equations to be manipulated (see Figure 7.1).

– *Phase III*: Solving the problem

The optimum of the problem is determined using optimization methods (see the Appendix on optimization).

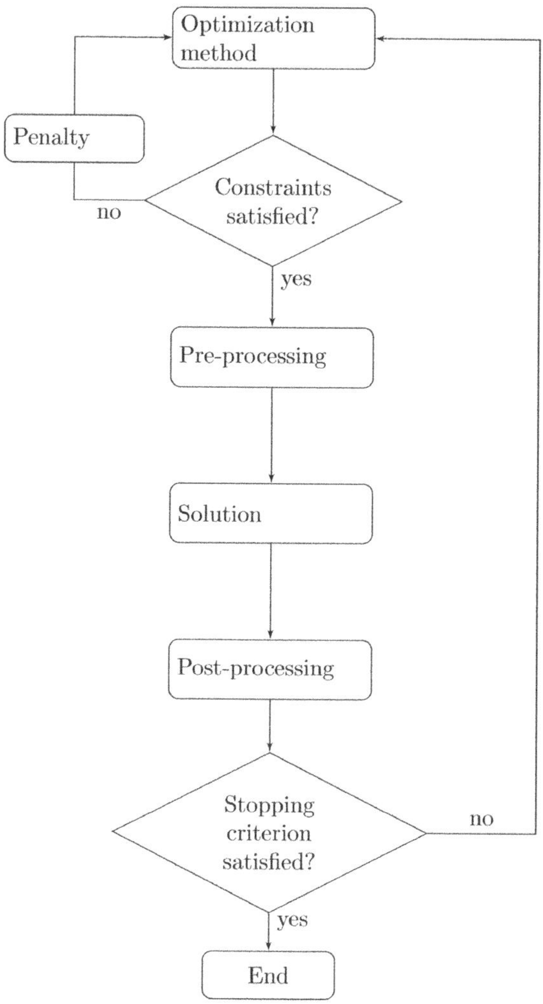

Figure 7.1. *Methodology for optimizing structures modeled by finite elements*

– *Phase IV*: Exploiting the solution

Once the problem has been solved, it is essential to evaluate the quality of the solution and review the various choices made at each step if the quality is insufficient. As noted above, the failure of an attempted optimization is often blamed on the search method used to find the optimum. But it is also essential to review the phases 1 and 2, since solving a poorly formulated problem (whether at the level of the specifications or the mathematical problem itself) can only ever produce an erroneous solution, no matter how perfectly a solving technique is used.

EXAMPLE 7.7.– Consider a steel plate experiencing a force $F = 10$ N/mm, as shown in Figure 7.2. The design variables P_1 and P_2 are also shown in the figure; they must take values between $16 < P_1, P_2 < 40$. Let us minimize the von Mises stress σ_{von} without exceeding the initial volume V_0.

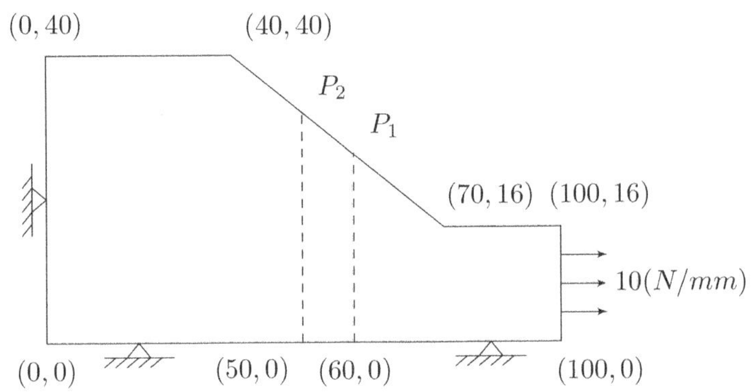

Figure 7.2. *Plate*

This is a shape optimization problem. It can be stated as follows:

$$\begin{cases} \min\ \sigma_{\text{von}} \\ \text{s.t.}\ \ V \leq V_0 \\ \qquad 16 \leq P_i \leq 40, \quad i = 1, 2 \end{cases} \qquad [7.38]$$

To solve this problem, we can call the *fmincon* method defined in the Matlab toolbox. The results of the calculations are given in Table 7.1.

Figure 7.3(b) shows the shape of the plate before optimization, and Figure 7.4 shows the optimal shape of the plate.

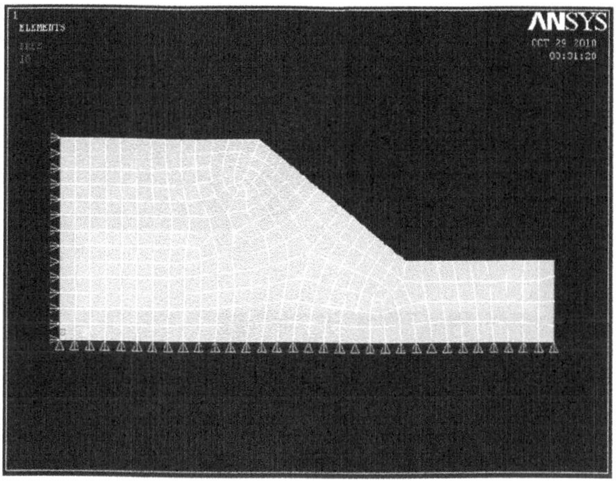

(a) Finite element model of the plate

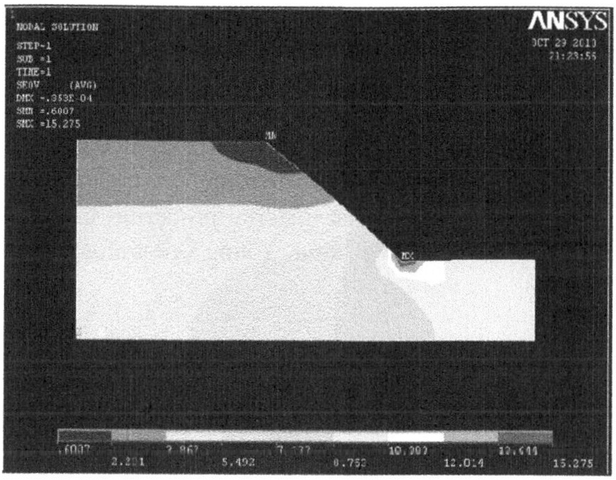

(b) Initial shape of the plate

Figure 7.3. *Shape of the plate. For a color version of this figure, see www.iste.co.uk/radi/optimizations.zip*

	Starting point	*fmincon*
P_1	24	17.99
P_2	32	22.91
σ	15	12.03
V	$2,920$	2746.41

Table 7.1. *Results calculated with Matlab*

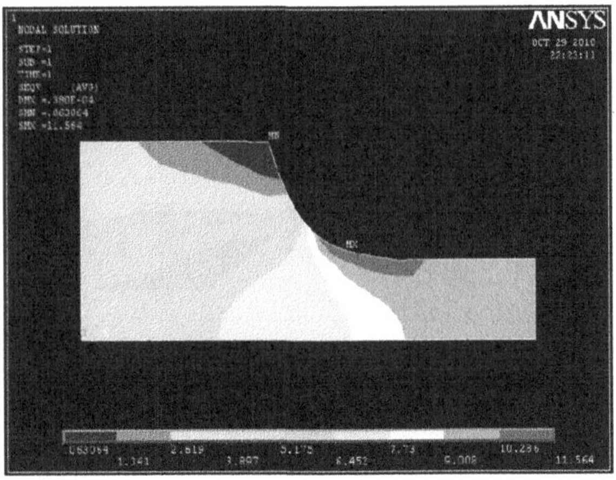

Figure 7.4. *Optimized plate. For a color version of this figure, see www.iste.co.uk/radi/optimizations.zip*

EXAMPLE 7.8.– Consider a triangular steel plate that can be fitted into various mechanical machines, as shown in Figure 7.5. The design variables are the thickness T_1 and the mesh radius FIL. A load corresponding to a traction pressure of 50 MPa is applied to the three planar faces. The material properties used in the analysis are specified as follows: thickness $T = 20$ mm, Young's modulus $E = 2.07e5$ MPa and Poisson's ratio $\nu = 0.3$.

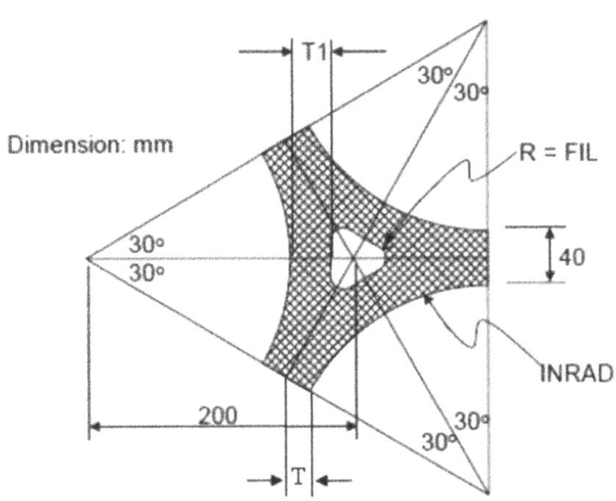

Figure 7.5. *Triangular plate*

We want to minimize the volume of the triangular plate V without exceeding the admissible von Mises stress of $\sigma_a = 150$ MPa.

This problem takes advantage of the symmetry of the geometry and the load applied to the structure. The problem can be formulated as follows:

$$\begin{cases} \min\ V \\ \text{s.t.}\ \ \sigma_{\text{von}} \leq \sigma_a \\ \qquad 20 \leq T_1 \leq 40, \\ \qquad 5 \leq FIL \leq 15. \end{cases} \quad [7.39]$$

The results are presented in Table 7.2.

	Initial point	Calculated solution
V	$22,967$	$17,952$
σ_{von}	56.73	149.86
T_1	30	20.0001
FIL	10	7.3160

Table 7.2. *Solution calculated with Matlab*

The initial shape of the triangular plate is shown in Figure 7.6 (a), and the optimal shape is shown in Figure 7.6 (b).

7.8. Using Matlab

The optimization toolbox offers various functions with a range of options for solving constrained optimization problems, for example:

– *fmincon*: minimizes a function of multiple constrained variables with various algorithms: trust region reflective algorithm, active set algorithm, SQP algorithm, interior point algorithm. The *fmincon* function is called with the syntax *[x,fval]= fmincon(fun,x0,A,b,Aeq,beq,lb,ub,nonlcon,options)* to specify inequality and equality constraints and bounds on the variable x. We can choose a specific algorithm in the options, for example:
options = optimoptions('fmincon','Algorithm','trust-region-reflective','Hessian', 'user-supplied').

– *fminbnd*: minimizes a scalar function of a bounded variable. This function has the following syntax: *[x,fval] = fminbnd(fun,x1,x2,options).*

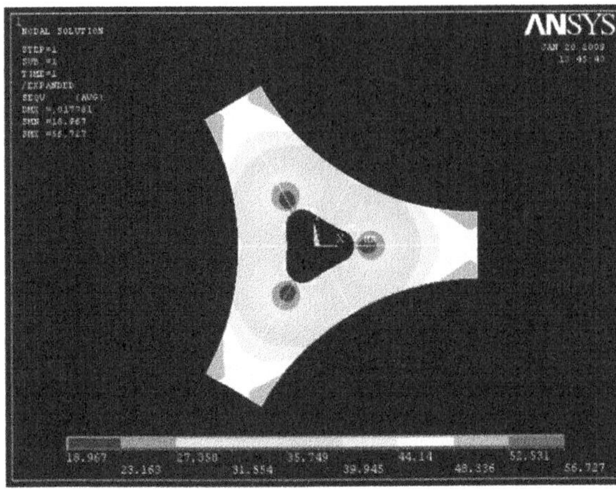

(a) Initial triangular plate

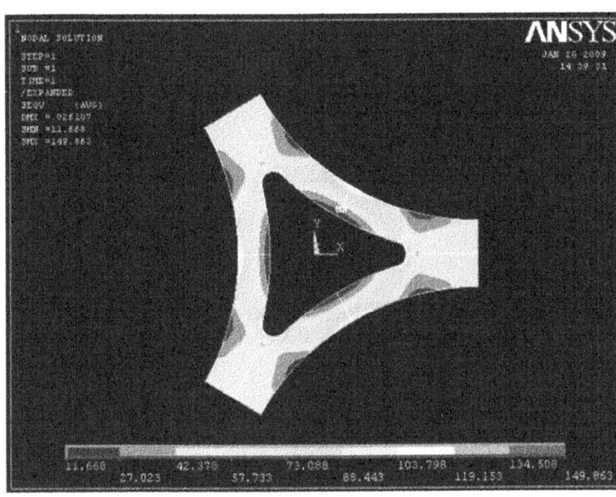

(b) Optimized triangular plate

Figure 7.6. *Initial and optimal shape. For a color version of this figure, see www.iste.co.uk/radi/optimizations.zip*

– *fseminf*: finds the minimum of the following problem:

$$\begin{cases} \min_x \ f(x) \\ \text{s.t.} \ \ Ax \leq b \\ \qquad A_{eq}x = b_{eq} \\ \qquad lb \leq x \leq ub \\ \qquad c(x) \leq 0 \\ \qquad ceq(x) = 0 \\ \qquad K_i(x, w_i) \leq 0, \ 1 \leq i \leq n. \end{cases} \tag{7.40}$$

The syntax is: *x = fseminf(fun,x0,ntheta,seminfcon,A,b,Aeq,beq,lb,ub,options).*

7.8.1. *The* fmincon *function*

The general topology of an optimization process with constrained variables can be written as follows in compact notation:

$$\begin{cases} \min_x \; f(x) \\ \text{s.t.} \;\; Ax \leq b \\ \qquad A_{eq}x = b_{eq} \\ \qquad c(x) \leq 0 \\ \qquad ceq(x) = 0 \\ \qquad lb \leq x \leq ub \end{cases} \qquad [7.41]$$

The quantities x, b, beq, lb and ub are vectors; A and Aeq are matrices; and $f(x)$, $c(x)$ and $ceq(x)$ are functions, possibly nonlinear. To solve a constrained optimization problem, we can use the *fmincon* function. This function is used to optimize multidimensional nonlinear objective functions. By default, the optimization algorithm is based on the SQP method. The standard syntax of this function is as follows: *[x,fval]= fmincon(fun,x0,A,b,Aeq,beq,lb,ub,nonlcon,options).* The *nonlcon* M-file function contains the nonlinear equality and inequality constraints. The other input and output arguments are the same as in the functions presented above. It is important to note that any missing arguments must be replaced by an empty set [ ]. The order of the input arguments is also important; we must always start with the inequality constraints, even if entering an empty set.

We can select one of the various algorithms in the options, for example: *options = optimoptions('fmincon','Algorithm','trust-region-reflective','Hessian', 'user-supplied').*

Let us begin by solving an optimization problem with linear constraints:

$$\begin{cases} \min\limits_{x \in \mathbb{R}^n} \; f(x_1, x_2) = 2x_1^2 + x_1x_2 + 2x_2^2 - 6x_1 - 6x_2 + 15 \\ \text{s.t.} \;\; x_1 + 2x_2 \leq 5 \\ \qquad 4x_1 \leq 7 \\ \qquad x_2 \leq 2 \\ \qquad -2x_1 + 2x_2 = -1 \end{cases} \qquad [7.42]$$

The Matlab script is as follows:

```
clc ;
clear all ;
%%%%%%%%%%%%%%%%%%%% Constrained optimization %%%%%%%%%%%%%%%%%%%%
fun = @(x) 2*x(1).^2 + x(1)*x(2) + 2*x(2).^2 - 6*x(1) - 6*x(2) + 15 ;
A = [1 2 ; 4 0 ; 0 1];
B = [5 7 2];
Aeq = [-2 2];
Beq = -1;
xinit = [-1 1/2];
options = optimset('LevenbergMarquardt','on','Display','iter',
          'TolX',1e-4);
[x, fval, exitflag, output, lambda, grad, hessian] =
                fmincon(fun,xinit, A, B, Aeq, Beq, [], [], [], options)
```

For a color version of this code, see www.iste.co.uk/radi/optimizations.zip

The optimization field *optimset('LevenbergMarquardt','on', ...)* stipulates that the optimization should be solved using the Levenberg–Marquardt algorithm. This algorithm can be alternatively selected with the syntax *optimset('NonlEqnAlgorithm','lm', ...)*. Similarly, we can use Gauss-Newton with the syntax *optimset('NonlEqnAlgorithm','gn', ...)*.

Executing this script yields the result after three iterations by calculating the gradient and the Hessian. The optimal solution and the value of the cost function are as follows:

```
x =

    1.4500    0.9500

fval =

    7.9875
```

Hence, the optimum is $x^* = (1.45, 0.95)^T$ with $f^* = 7.9875$.

7.8.2. *The* fminbnd *function*

Optimization methods for functions of a single variable are known as "line search". The algorithms implemented by Matlab for the *fminbnd* function are "Golden section search" and parabolic interpolation.

We will create an external function in an .m file because working with external files allows us to simplify the code and avoid errors. Thus, let us minimize the function $f(x) = \frac{1}{(x-0.3)^2+0.01} + \frac{1}{(x-0.9)^2-6}$, defined in a file named "f.m" as follows:

```
function y = f(x)
y = 1./((x-0.3).^2 + 0.01)+ 1./((x - 0.9).^2 + 0.04) -6;
```

For a color version of this code, see www.iste.co.uk/radi/optimizations.zip

By observing the graph of the function f, we can see that the minimum is found between $0.3 \leq x \leq 1$, so we will call the function *fminbnd* on the interval $[0.3, 1]$:

```
min = fminbnd('f',0.3,1,optimset('Display','iter'));
fplot('f',[0 2])
hold on ;
plot(min,f(min),'r*') ;
```

For a color version of this code, see www.iste.co.uk/radi/optimizations.zip

In the options, we can see the successive approximations and the algorithm used by Matlab with optimset(*'Display'*, *'iter'*). The function *fminbnd* finds local minima. It is important to choose a good initial approximation. It converges slowly when the solution is close to the interval. The final iterations with a precision of 1.e-04 are presented below:

```
8    0.637019    11.2528    parabolic
9    0.637052    11.2528    parabolic
```

Hence, the optimal solution is $x^* = 0.637052$ with a cost of $f(x^*) = 11.2528$. Figure 7.7 presents the optimum calculated by the algorithm. We can always try changing the search interval to investigate the effect on the speed of convergence.

7.8.3. *Penalty method*

Consider the following minimization problem:

$$\begin{cases} \min\ f(x_1, x_2) = 2x_1 + x_2 + (x_1^2 - x_2^2) + (x_1 - x_2^2)^2 \\ \text{s.t.}\ \ g(x_1, x_2) = 8x_1 - 6x_2 + 21 \leq 0 \end{cases} \qquad [7.43]$$

Figure 7.8 presents the graphical representations of the functions f and g. Figure 7.9 shows the contours of the functions f and g.

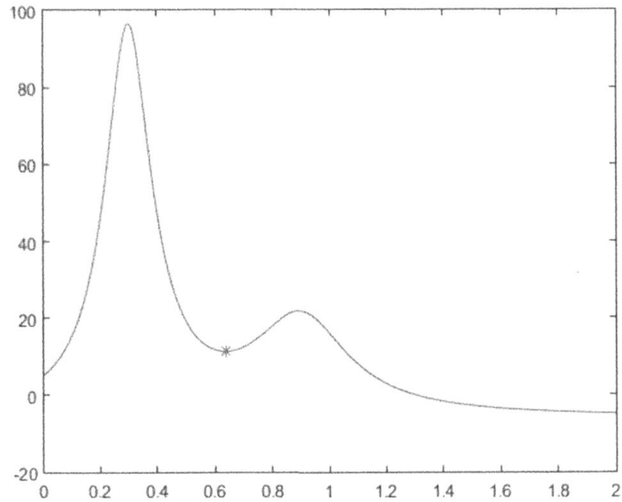

Figure 7.7. *Position of the local optimum. For a color version of this figure, see www.iste.co.uk/radi/optimizations.zip*

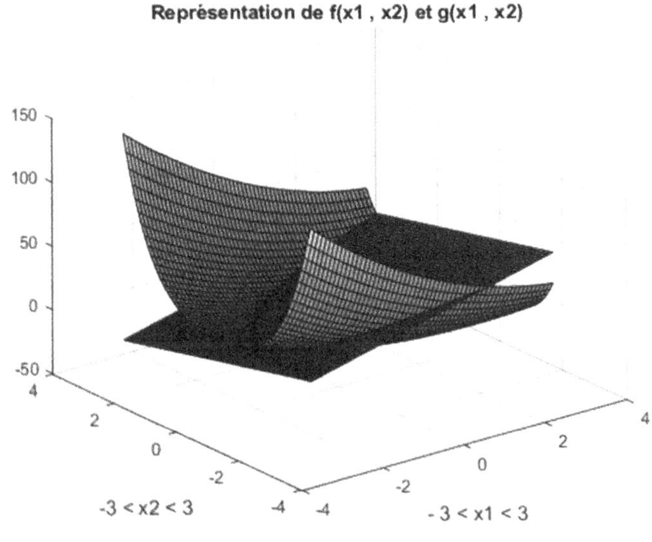

Figure 7.8. *Graphs of the functions f and g. For a color version of this figure, see www.iste.co.uk/radi/optimizations.zip*

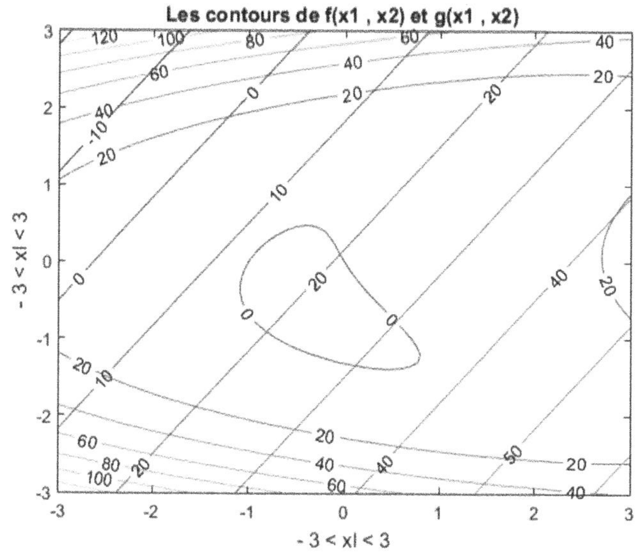

Figure 7.9. *Contours of the functions f and g. For a color version of this figure, see www.iste.co.uk/radi/optimizations.zip*

These graphs are generated by the following script:

```
%
% Graph of the objective function and
% the inequality constraint
%
% Construct the axes
[x1,x2] = meshgrid(-3:0.1:3,-3:0.1:3);
%
% Objective function
f = 2*x1+x2+(x1.^2-x2.^2)+(x1-x2.^2).^2;
%
% Constraint
g = 8*x1-6*x2+21;
%
% Construct f (x1 , x2) and g (x1 , x2)
subplot(2,1,1);
figure
surf(x1,x2,f);
hold on;
%
surf(x1,x2,g);
hold off;
title('Representation of f(x1 , x2) and g(x1 , x2)');
xlabel('- 3 < x1 < 3');
```

```
ylabel('-3 < x2 < 3') ;
% Contours of f(x1 , x2) and g(x1 , x2)
subplot (2, 1, 2);
figure
contour(x1,x2,f,'ShowText','on')
hold on;
contour(x1,x2,g,'ShowText','on')
hold off;
title('The contours of f(x1 , x2) and g(x1 , x2)');
xlabel('- 3 < x1 < 3');
ylabel ('- 3 < x1 < 3');
```

For a color version of this code, see www.iste.co.uk/radi/optimizations.zip

The following Matlab script finds the optimum using the exterior penalty method.

```
%
% Exterior penalty method
% based on the gradient method
%
% Assumptions:
x1zero = -3.0;    % starting value of x1
x2zero = 3.0;     % starting value of x2
epsilon = 1e-3;   % precision
alfa0 = 0.001;    % initial jump factor
incAlfa = 0.001; % jump factor increment
Rp = 1.0;         % penalty factor
%
incRp = 0.5; % increment
Δ = 0.00001;
% initialization
ΔF = 1e3; ΔFp = 1e3;
%
%
fp = fopen( ' ExtPenalty.txt ','w');
%
fprintf (fp, ' Exterior penalty method\n ');
fprintf (fp, ' based on the gradient method ');
fprintf (fp, ' implementation\n\n ');
fprintf (fp, ' initial point X0(x1zero,x2zero) : \n ');
fprintf (fp, ' x1zero = %6.3f x2zero = %6.3f\n ',x1zero,x2zero);
fprintf (fp, ' precision: epsilon = %f\n ', epsilon);
fprintf (fp, ' initial jump factor alfa0 = %f\n ',alfa0) ;
fprintf (fp, ' jump factor increment incAlfa = %f\n ',incAlfa);
fprintf (fp, ' penalty factor Rp = %f\n',Rp);
%
f_old - f(x1zero,x2zero);
fMin = f_old; fp1 = f_old;
while (ΔFp ≥ epsilon)
% direction of minimization
x1 = x1zero; x2 = x2zero;
if(g1(x1,x2) ≥ 0)
```

```
    fa = pseudo_f(x1,x2,Rp);
else
    fa = f(x1,x2);
end
%
x1 = x1 + Δ;
if (g1(x1,x2) ≥ 0)
    fb = pseudo_f(x1,x2,Rp);
else
    fb = f(x1,x2);
end
diff = fa − fb; s1= −diff/(−Δ); x2= x2 + Δ;
if (g1(x1,x2) ≥ 0)
    fb = pseudo_f(x1,x2,Rp);
else
    fb = f(x1,x2);
end
%
diff = fa − fb; s2 = −diff/(−Δ);
% compute x1
alfa = alfa0;
x1one = x1zero + alfa*s1;
x2one = x2zero + alfa*s2;
if (g1(x1one,x2one) ≥ 0)
    f_new = pseudo_f(x1one,x2one,Rp);
else
    f_new = f(x1one,x2one);
end
%
while(f_new < f_old)
    f_old = f_new;
    alfa = alfa + incAlfa;
    x1unu = x1zero + alfa*s1;
    x2unu = x2zero + alfa*s2;
    if(g1(x1one,x2one) ≥ 0)
        f_new= pseudo_f(x1one,x2one,Rp);
    else
        f_new=f(x1one,x2one);
    end
end
%
fMin = f_new; x1Min = x1one; x2Min = x2one;
ΔF = abs(f_old − f_new);
%
while(ΔF ≥ epsilon)
    func = f_new;
    x1p = x1Min;
    x2p = x2Min;
%
% direction of minimization
    x1 = x1p; x2 = x2p;
    if(g1(x1,x2) ≥ 0)
        fa = pseudo_f(x1,x2,Rp);
    else
        fa = f(x1,x2);
```

```
        end
%
        x1 = x1 + Δ;
        if (g1(x1,x2) ≥ 0)
            fb = pseudo_f(x1,x2,Rp);
        else
            fb = f(x1,x2);
        end
%
        diff = fa − fb; s1 = −diff/(−Δ); x2 = x2 + Δ;
%
        if (g1(x1,x2) ≥ 0)
            fb = pseudo_f(x1,x2,Rp);
        else
            fb=f(x1,x2);
        end
        diff = fa − fb; s2 = −diff/(−Δ);
% compute x1
        alfa = alfa0; x1one = x1p + alfa*s1; x2one = x2p + alfa*s2;
        if (g1(x1one,x2one ≥ 0))
            f_new = pseudo_f(x1one,x2one,Rp);
        else
            f_new = f(x1one,x2one);
        end
%
        while(f_new < f_old)
            f_old = f_new; alfa = alfa + incAlfa;
            x1unu = x1p + alfa*s1; x2unu = x2p + alfa*s2;
            if(g1(x1one, x2one ≥0)
                f_new = pseudo_f(x1one,x2one,Rp);
            else
                f_new = f(x1one,x2one);
            end
        end
        fMin = f_new; x1Min = x1one; x2Min = x2one;
        ΔF = abs(func−f_new);
end
%
        fp2 = fMin; ΔFp = abs(fp1 − fp2);
        fp1 = fp2; Rp = Rp + incRp;
end
% save the final result
fprintf (fp, ' \n Final result: \n ');
fprintf (fp, ' x1Min = %f x2Min = %f \n fMin =%f\n ',x1Min,x2Min,
      f(x1Min,x2Min));
fprintf (fp, ' constraint g1 = %f\n ',g1(x1Min, x2Min));
% close the file ' ExtPenalty.txt '
fclose(fp);
```

For a color version of this code, see www.iste.co.uk/radi/optimizations.zip

To execute this script, we need to call the functions "f.m", "g1.m" and "pseudo_f.m". We find a minimum point with coordinates $(-2.125745, 0.657550)$, and a cost function of 7.036444.

```
function [val] = f(x1,x2)
val = 2*x1+x2+(x1.^2-x2.^2)+(x1 - x2.^2).^2;
```

For a color version of this code, see www.iste.co.uk/radi/optimizations.zip

```
function [val] = g1(x1,x2)
val = 8*x1-6*x2+21;
return;
end
```

For a color version of this code, see www.iste.co.uk/radi/optimizations.zip

and

```
function [val] = pseudo_f(x1,x2,Rp)
val = f(x1,x2)+Rp*g(x1,x2)*g(x1,x2);
return;
end
```

For a color version of this code, see www.iste.co.uk/radi/optimizations.zip

For more details, the reader can refer to [ANC 19].

Appendices

Appendix 1

Reminders from Linear Algebra

$\mathbb{R}$ and $\mathbb{C}$ denote the field of real numbers and the field of complex numbers, respectively. If there is no need to distinguish between them, we instead simply speak of the field $\mathbb{K}$ of scalars. The set E equipped with the two laws of addition and multiplication by a scalar denotes a vector space over $\mathbb{K}$ (or $\mathbb{K}$-vector space).

A1.1. Vector space

A1.1.1. *General definitions*

DEFINITION A1.1.– *A vector space over the field $\mathbb{K}$ is a set E equipped with the following two laws:*

– *addition, giving E the structure of a commutative group;*

– *the "exterior" product of an element of E by an element of $\mathbb{K}$, with the following properties:*

- $\forall \lambda, \mu \in \mathbb{K}, \forall x \in E : (\lambda\mu)x = \lambda(\mu x)$
- $\forall \lambda, \mu \in \mathbb{K}, \forall x \in E : (\lambda + \mu)x = \lambda x + \mu x$
- $\forall \lambda \in \mathbb{K}, \lambda(x + y) = \lambda x + \lambda y$
- $\forall x \in E : 1x = x$ *(where* 1 *is the identity element of* $\mathbb{K}$*).*

The elements of E are called vectors.

DEFINITION A1.2.– *Let F be a subset of the vector space E. We say that F is a vector subspace of E if it is fixed under the operations of E; in other words:*

– $\forall x, y \in F, x + y \in F$;

– $\forall \lambda \in \mathbb{K}, \forall x \in F, \lambda x \in F$.

DEFINITION A1.3.– *Let E be a $\mathbb{K}$-vector space, and let F and G be two vector subspaces of E.*

1) The sum of F and G, denoted $F + G$, is the set

$$F + G = \{x = y + z,\ y \in F,\ z \in G\}.$$

This is a vector subspace.

2) F and G are said to form a direct sum if $F \cap G = \{0\}$. If so, we write this sum as $F \oplus G$.

3) If we also have $E = F \oplus G$, we say that F and G are supplementary subspaces.

THEOREM A1.1.– If F and G are supplementary, then, for any element x of E, there exists a unique pair (y, z) in $F \times G$ such that $x = y + z$.

A1.1.2. *Free families, generating families and bases*

DEFINITION A1.4.– *Let $B = \{x_1, \dots, x_p\}$ be a family of vectors of E.*

– We say that B is related if one of its vectors is a linear combination of the others, namely:

$$\exists (t_1, \dots, t_p) \in \mathbb{K}^p,\ (t_1, \dots, t_p) \neq (0, \dots, 0) \textit{ such that } \sum_{i=1}^{p} t_i x_i = 0.$$

– We say that B is free if it is not related. If so, its vectors are linearly independent.

– We say that B is a generator of E (or generates E) if every element of E is a linear combination of the elements of B.

DEFINITION A1.5.– *A family $B = \{e_1, \dots, e_p\}$ of elements of the vector space E is a basis of E if it is free and generates E.*

There is a particular basis called the canonical basis that is defined as follows:

DEFINITION A1.6.– *The canonical basis is the basis formed by the vectors $\{e_i\}_{i=1\dots n}$ such that the jth element of $\vec{e}_i$ is 0 except when $i = j$, in which case it is 1.*

Thus, every vector x in $\mathbb{R}^n$ decomposes as follows with respect to the canonical basis:

$$\forall x \in \mathbb{R}^n,\ x = \sum_{i=1}^{n} x_i e_i.$$

DEFINITION A1.7.– *In a vector space generated by a finite family, every basis has the same number of elements.*

DEFINITION A1.8.– *The number of elements of any basis of E is said to be the dimension of a vector space E generated by a finite family, denoted "dim E".*

If we always work with the same basis $B = \{e_1, \dots, e_n\}$ in a vector space with finite dimension n, then any vector x in E decomposes uniquely as follows with respect to B:

$$x = x_1 e_1 + \cdots + x_n e_n.$$

The element $X = (x_1, \dots, x_n)$ of $\mathbb{K}^n$ can be chosen as a representation of x without any ambiguity.

THEOREM A1.2.– Let E be an n-dimensional vector space, and let F and G be two vector subspaces of E. Then:

1) every free family of n vectors is a basis;

2) every generating family of n vectors is a basis;

3) dim $F \leq$ dim E;

4) if $F \cap G = \{0\}$, then dim $F+$ dim $G \leq$ dim E;

5) if also $E = F \oplus G$, then dim $F+$ dim $G =$ dim E.

A1.2. Linear mappings

DEFINITION A1.9.– *Let E and F be two vector spaces over the field K. A mapping $u\colon E \to F$ is said to be a linear mapping if it satisfies the following properties:*

– $u(x + y) = u(x) + u(y)\ \forall x, y \in E$;

– $u(\lambda x) = \lambda u(x)\ \forall x \in E,\ \forall \lambda \in K$.

The set of linear mappings from E into F is denoted $\mathcal{L}(E, F)$.

Note that linear mappings are mappings that conserve the vector space structure.

DEFINITION A1.10.–

1) The kernel of u, denoted Ker(u), is the vector subspace of E defined by:

$$\mathit{Ker}(u) = \{x \in E | u(x) = 0\}.$$

2) The image of u, denoted Im(u), is the vector subspace of F defined by:

$$\mathit{Im}(u) = \{y \in F \mid \exists x \in E \text{ such that } y = u(x)\}.$$

THEOREM A1.3.–

– u is injective if and only if Ker$(u) = \{0\}$;

– u is surjective if and only if Im$(u) = F$.

DEFINITION A1.11.– *Let I_E (respectively, I_F) be the identity mapping of E (respectively, F). The linear mapping u from E into F is invertible if there exists a linear mapping u^{-1} from F into E such that:*

$$u^{-1} \circ u = i_E \quad \textit{and} \quad u \circ u^{-1} = i_F. \qquad [A1.1]$$

Any invertible linear mapping is therefore bijective; in other words, it is injective and surjective.

THEOREM A1.4.– Let $u \in \mathcal{L}(E, E)$. The following are equivalent:

– u is injective;

– u is surjective;

– u is bijective.

THEOREM A1.5.– Let $u \in \mathcal{L}(E, F)$, and let $B = \{e_1, \dots, e_n\}$ be a basis of E. Then:

– if u is injective, $\{u(e_1), \dots, u(e_n)\}$ is a basis of Im(u);

– if u is surjective, $\{u(e_1), \dots, u(e_n)\}$ is a generating family of F;

– we have the following relation:

$$\dim E = \dim \text{Ker}(u) + \dim \text{Im}(u).$$

DEFINITION A1.12.– *The rank of a linear mapping, denoted rank u, is called the dimension of Im(u).*

A1.3. Matrices

In this section, E, F and G represent three vector spaces over the field $\mathbb{K}$ with finite dimensions n, p and q, respectively. The families $B_E = \{e_1, \dots, e_n\}$, $B_F = \{f_1, \dots, f_p\}$ and $B_G = \{g_1, \dots, g_q\}$ are bases of E, F and G.

DEFINITION A1.13.– *Matrix. Let $u \in \mathcal{L}(E, F)$. The matrix associated with u with respect to the bases B_E and B_F is the table A of scalars (i.e. of elements of $\mathbb{K}$) with p rows and n columns such that the jth column of A is given by the components of the vector $u(e_j)$ with respect to the basis B_F.* If a_{ij} is the element of A in the ith row and the jth column, then:

$$u(e_j) = \sum_{i=1}^{p} a_{ij} f_i.$$

A matrix A with p rows and n columns is said to be a $p \times n$ matrix, or a matrix of type (p, n).

We cannot reasonably prove anything about matrices without referring to the linear mappings that they can represent. This connection is used to define operations on matrices.

A1.3.1. *Operations on matrices*

DEFINITION A1.14.– *Let A and B be two matrices, both of type (p, n). The sum $A + B$ of A and B is defined as the matrix C of type (p, n) whose coefficients c_{ij} are given by*

$$c_{ij} = a_{ij} + b_{ij}.$$

The matrix C is the matrix of the linear mapping formed by the sum of the linear mappings represented by A and B. Similarly, we define the product of the matrix A by the scalar λ as the matrix λA obtained by multiplying each coefficient of A by λ. The set of matrices with p rows and n columns forms a vector space denoted $\mathcal{M}_{p,n}(\mathbb{K})$ (or $\mathcal{M}_{p,n}$ if the base field is implicit).

DEFINITION A1.15.– *Let $u \in \mathcal{L}(F, G)$ and $v \in \mathcal{L}(E, F)$. Set $w = u \circ v$ (and therefore $v \in \mathcal{L}(E, G)$). Let A be the matrix of u with respect to the bases B_F and B_G, and let B be the matrix of v with respect to the bases B_E and B_F. The matrix of w with respect to the bases B_E and B_G is defined as the product of A and B, denoted AB.*

THEOREM A1.6.– Let $A \in \mathcal{M}_{n,p}$ and $B \in \mathcal{M}_{p,q}$. Let $C = AB$. Then $C \in \mathcal{M}_{n,q}$ and its elements are given by the formula:

$$c_{ij} = \sum_{k=1}^{p} a_{ik} b_{kj} \quad i = 1, \ldots, n, \; j = 1, \ldots, q. \qquad [A1.2]$$

DEFINITION A1.16.– *A matrix A is invertible (also said to be regular or non-singular) if the linear mapping associated with it is invertible. If $A \in \mathcal{M}_{n,n}$ is invertible, then there exists an inverse matrix, denoted A^{-1}, such that*

$$AA^{-1} = A^{-1}A = I,$$

where I is the identity matrix of $\mathcal{M}_{n,n}$.

Properties

Assuming that A, B and C have suitable dimensions for their products to be defined, for $\alpha \in \mathbb{R}$, we have:

- $C(A+B) = CA + CB$;

- $(A+B)C = AC + BC$;

- $A(BC) = (AB)C = ABC$;

- $\alpha(AB) = (\alpha A)B = A(\alpha B)$.

A1.3.2. *Change of basis matrices*

DEFINITION A1.17.– *Let E be an n-dimensional vector space equipped with a basis $B_E = \{e_1, \ldots, e_n\}$. Consider another basis of E, denoted $B' = \{e'_1, \ldots, e'_n\}$. The change of basis matrix P is the matrix whose columns are given by the components of the elements of E with respect to the first basis. Thus:*

$$e'_j = \sum_{i=1}^{n} p_{ij} e_i.$$

The change of basis matrix does not allow us to compute the new components of a vector $\vec{x}$ directly in terms of the old components. Set

$$x = \sum_{i=1}^{n} x_i e_i = \sum_{i=1}^{n} x'_i e'_i.$$

Let X be the vector of K^n formed by the components of x with respect to the old basis B_E, and let X' be the vector of $\mathbb{K}^n$ formed by the components of x with respect to the new basis B'_E. It can be shown that:

$$X = PX'.$$

In other words, given that a change of basis matrix is always invertible:

$$X' = P^{-1}X.$$

THEOREM A1.7.– Let $u \in \mathcal{L}(E, E)$, with matrix A with respect to the basis B_E. The matrix A' of u with respect to the basis B'_E is then given by the formula

$$A' = P^{-1}AP.$$

Change of basis operations are fundamental in practice because they allow us to express square matrices in more "pleasant" forms (diagonal, upper triangular or lower triangular structure).

A1.3.3. *Matrix notation*

Unless otherwise stated, given $A \in \mathcal{M}_{p,n}$, we shall adopt the following notation:

– A_j denotes the jth column of the matrix A, which can therefore be represented as:

$$A = (A_1, A_2, \ldots, A_n).$$

– A^T denotes the transpose of the matrix A (obtained by exchanging the rows and columns of A). We therefore have $A^T \in \mathcal{M}_{n,p}$ and:

$$(A^T)_{ij} = a_{ji} \quad \forall i, j,\ 1 \leq i \leq p, 1 \leq j \leq n.$$

– If $A \in \mathcal{M}_{np}(\mathbb{C})$, we write $\bar{A}$ for the complex conjugate matrix of A (whose coefficients are the complex conjugates of the coefficients of A).

– If A is a square matrix ($p = n$), we define the trace of A as the sum of its diagonal elements:

$$\text{tr}(A) = \sum_{i=1}^{n} a_{ii}.$$

– We write $D = \text{diag}(d_1, \ldots, d_n)$ for the diagonal matrix D.

– Every vector of K^n is identified with a matrix with n rows and one column. If X is a column vector, X^T represents a row vector (matrix with one row).

DEFINITION A1.18.– *A matrix $A \in \mathcal{M}_{n,n}$ that satisfies*

$$a_{ij} = a_{ji} \quad \forall i, j \in \{1, \ldots, n\}$$

is called a symmetric matrix. Any such matrix satisfies $A^T = A$.

A1.4. Determinants

The definition of the determinant requires us to introduce the notions of permutation and signature, which are presented below.

DEFINITION A1.19.– *Let $I_n = \{1, 2, \ldots, n\}$. A permutation is a bijective mapping from I_n onto itself. A transposition is a permutation that only permutes two consecutive elements (τ_j represents the permutation that exchanges j and $j+1$).*

We shall write $\mathcal{P}_n$ for the set of permutations of I_n. Every permutation can be expressed as a product of transpositions. This decomposition is not unique, but the parity of the number of transpositions is always constant.

DEFINITION A1.20.– *The signature of a permutation σ is a number, denoted sign(σ), that is equal to +1 if the permutation decomposes into evenly many transpositions, and -1 otherwise.*

DEFINITION A1.21.– *The determinant of a matrix $A \in \mathcal{M}_{nn}$, denoted det(A), is the number*

$$det(A) = \sum_{\sigma \in \mathcal{P}} sign(\sigma) a_{\sigma(1),1} a_{\sigma(2),2} \cdots a_{\sigma(n),n}. \qquad [A1.3]$$

The number of permutations of a finite set is finite, and the cardinality of $\mathcal{P}_n$ is finite. Therefore, the sum in the formula [A1.3] is a finite sum.

Properties

The key properties of the determinant are as follows:

1) $\det(\bar{A}) = \overline{\det(A)}$;

2) $\det(A) = \det(A^T)$;

3) $\forall \sigma \in \mathcal{P}_n$, $\det(A_{\sigma(1)}, A_{\sigma(2)}, \ldots, A_{\sigma(n)}) = \text{sign}(\sigma)\det(A)$;

4) the mapping that sends a column of A to the determinant of A is linear. In particular, if a column of A is a linear combination of the other columns, then $\det(A) = 0$.

The mapping that sends the columns of the matrix A to the determinant of A is a multilinear alternating mapping. Multilinearity follows from property 4. This implies that

$$\det(\alpha A) = \alpha^n \det(A).$$

It is an alternating mapping in the sense that, by property 3, permuting any two consecutive columns of the matrix changes the sign of the determinant.

Properties

1) In general, $\det(A + B) \neq \det(A) + \det(B)$.

2) Let A and B be two matrices of $\mathcal{M}_{nn}$. Then:

$$\det(AB) - \det(A)\det(B).$$

Let $A \in \mathcal{M}_{nn}(K)$. Write $A_{|i,j|} \in \mathcal{M}_{n-1,n-1}(K)$ for the minor submatrix of A obtained by deleting the ith row and the jth column. From a practical perspective, determinants are usually computed using the following properties:

THEOREM A1.8.– Developing along a column. We can calculate the determinant by developing along a column:

$$\forall j \in \{1, \ldots, n\} \quad \det(A) = \sum_{i=1}^{n} a_{ij}(-1)^{i+j}\det(A_{|i,j|}). \qquad [A1.4]$$

We can also develop along a row with an analogous formula by property 2 above.

DEFINITION A1.22.– *The scalar* $(-1)^{i+j}det(A_{|i,j|})$ *is called the cofactor of the element* a_{ij}. *The matrix of cofactors is called the comatrix and is denoted* $co(A)$.

THEOREM A1.9.– A square matrix A is invertible (or regular or non-singular) if and only if its determinant is non-zero. If so, it satisfies the following inversion formula:

$$A^{-1} = \frac{1}{det(A)}[co(A)]^T. \qquad [A1.5]$$

A1.5. Scalar product

DEFINITION A1.23.– *A scalar product on* E *is a mapping from* $E \times E$ *onto* $\mathbb{R}$*:*

$$(\vec{x}, \vec{y}) \rightarrow < \vec{x}, \vec{y} >= (\vec{x}, \vec{y})_E$$

satisfying the following properties:

– $< \vec{x}, \vec{x} > \geq 0$;

– $< \vec{x}, \vec{x} >= 0 \Rightarrow \vec{x} = 0$;

– $< \vec{x}, \vec{y} >=< \vec{y}, \vec{x} >$;

– $< \vec{x}, \vec{y} + \vec{z} >=< \vec{x}, \vec{y} > + < \vec{x}, \vec{z} >$;

– $< \vec{x}, \alpha\vec{y} >= \alpha < \vec{x}, \vec{y} > \forall \alpha \in \mathbb{R}$.

THEOREM A1.10.– The scalar product is a symmetric bilinear form on E.

A1.6. Vector norm

There are several ways to define the norm of a vector. The most widely known is undoubtedly the Euclidean norm of a vector $\vec{x} = (x_1, x_2, \ldots, x_n)$:

$$\|(x_1, x_2, \ldots, x_n)\| = \sqrt{< \vec{x}, \vec{x} >} = \sqrt{\sum_{i=1}^{n} x_i^2}, \quad x_i \in \mathbb{R}.$$

Geometrically, this norm represents the length of the vector. Norms can, however, be defined more generally.

DEFINITION A1.24.– *A mapping* $\|.\| : E \to \mathbb{R}$ *(here,* E *is a real vector space) is said to be a vector norm if it satisfies the following properties:*

1) $\| \vec{x} \| \geq 0, \forall \vec{x} \in E$ *and* $\| \vec{x} \| = 0 \Leftrightarrow \vec{x} = 0$;

2) $\| \lambda \vec{x} \| = | \lambda | \| \vec{x} \|, \forall a \in \mathbb{R}, \vec{x} \in E$;

3) $\| \vec{x} + \vec{y} \| \leq \| \vec{x} \| + \| \vec{y} \|, \forall \vec{x}, \vec{y} \in E$.

It is easy to check that the definition of the Euclidean norm satisfies these properties. As noted above, several different norms can be defined on the same vector space. If we restrict attention to norms defined on $\mathbb{R}^n$, we can cite the so-called standard p-norms of a vector $\vec{x}$, which are calculated as follows:

$$\| \vec{x} \|_p = \left(\sum_{i=1}^{n} |x_i^p| \right)^{1/p} .$$

When $p = 2$, we recover the definition of the Euclidean norm. There is also the so-called infinity norm, which is defined as follows:

$$\|\vec{x}\|_\infty = \max_{1 \leq i \leq n} |x_i|.$$

The infinity norm, the 2-norm and the 1-norm are the most widely used. Slightly more exotic norms are also possible (e.g. elliptic norms), but we will work with the norms stated above.

EXAMPLE A1.1.– *Let us compute the* 1, 2 *and* ∞ *norms of the vector* $\vec{x} = (1, 0, 1, 4)^T$. *We find:*

$$\|\vec{x}\|_1 = \sum_{i=1}^{4} |x_i| = 6;$$

$$\|\vec{x}\|_2 = \sqrt{\sum_{i=1}^{4} x_i^2} = \sqrt{18} = 3\sqrt{2};$$

$$\|\vec{x}\|_\infty = \max_{1 \leq i \leq 4} | x_i | = 4.$$

As shown by the example, different norms do not necessarily lead to the same answer.

Appendix 2

Reminders about functions from $\mathbb{R}^n$ into $\mathbb{R}$

Notation:

– $x \in \mathbb{R}^n$ represents a column vector whose components are denoted $x_i, i = 1, \ldots, n$.

– We shall consider the space $\mathbb{R}^n$ equipped with the Euclidean scalar product

$$< x|y >= \sum_{i=1}^{n} x_i y_i$$

and the Euclidean norm

$$\|x\| = \sqrt{< x|x >} = \left(\sum_{i=1}^{n} x_i^2\right)^{\frac{1}{2}}.$$

A2.1. Differentiability

DEFINITION A2.1.– *Let $f\colon \mathbb{R}^n \to \mathbb{R}$. We say that f is continuous at the point x_0 if*

$$\forall \epsilon > 0,\ \exists \eta > 0 / \ \forall x \in \mathbb{R}^n,\ \|x - x_0\| \leq \eta \Rightarrow |f(x) - f(x_0)| \leq \epsilon.$$

If f is continuous at x_0 for every $x_0 \in \mathbb{R}^n$, we say that f is continuous on $\mathbb{R}^n$.

DEFINITION A2.2.– *Let $f\colon \mathbb{R}^n \to \mathbb{R}$. We say that f is differentiable at the point $x \in \mathbb{R}^n$ if there exists a vector $g(x) \in \mathbb{R}^n$ such that $h \in \mathbb{R}^n$:*

$$f(x + h) = f(x) + < g(x) \mid h > + \|h\|\epsilon(\|h\|),$$

where $\epsilon\colon \mathbb{R} \to \mathbb{R}$ is a continuous function such that $\lim_{\lambda \to 0} \epsilon(\lambda) = 0$.

Notation:

By convention, the gradient vector of f at x is a column vector $g(x)$ or $\overrightarrow{\text{grad}}f(x)$.

REMARK.– In general, in mathematics books, the **gradient** of f at the point x is the row vector $\nabla f(x)$ such that

$$\nabla f(x) = g(x).$$

With this notation, differentiability can be expressed in the form

$$f(x+h) = f(x) + \nabla f(x)^T.h + \|h\|\epsilon(\|h\|),$$

where $\nabla f(x).h$ represents the scalar product.

DEFINITION A2.3.– *Consider the (possibly non-continuous) function $f\colon \mathbb{R}^n \to \mathbb{R}$. Let x, $y \in \mathbb{R}^n$. We say that f admits a directional derivative at the point x in the direction y if the following limit exists:*

$$\lim_{\lambda \to 0, \lambda \neq 0} \frac{f(x+\lambda y) - f(x)}{\lambda}.$$

Notation:

The limit, if it exists, is called the directional derivative of f at x in the direction y, denoted $Df(x;y)$.

REMARK.– If f is differentiable at x, then it admits directional derivatives in every direction at x. The partial derivatives are just special cases of the directional derivative.

THEOREM A2.1.– Let f be a differentiable function with respect to $x \in \mathbb{R}^n$ and $y \in \mathbb{R}^n$ an arbitrary vector. Then:

$$\text{Df}(x;y) = \nabla f(x).y.$$

THEOREM A2.2.– The gradient vector is the vector of partial derivatives:

$$\forall i \in \{1,\ldots,n\}, \quad g_i(x) = \frac{\partial f}{\partial x_i}(x).$$

DEFINITION A2.4.– *The Hessian matrix (or Hessian) of f at x is the matrix $\nabla^2 f(x)$ whose elements are the second-order partial derivatives of the function f at the point x:*

$$\left[\nabla^2 f(x)\right]_{ij} = \frac{\partial^2 f}{\partial x_i \partial x_j}(x).$$

In the following, except where explicitly stated otherwise, we shall consider twice continuously differentiable functions whose Hessian matrices are always defined and symmetric.

A2.2. Convexity

DEFINITION A2.5.– *We say that* $f\colon \mathbb{R}^n \to \mathbb{R}$ *is convex if* $\forall x, y \in \mathbb{R}^n$ *and* $\forall \theta \in [0, 1]$,

$$f(\theta x + (1-\theta)y) \leq \theta f(x) + (1-\theta)f(y).$$

The function f *is strictly convex if, for all* $x, y \in \mathbb{R}^n$, $x \neq y$, *and for every* $\theta \in]0, 1[$,

$$f(\theta x + (1-\theta)y) < \theta f(x) + (1-\theta)f(y).$$

REMARK.– Let $f\colon \mathbb{R} \to \mathbb{R}$. The following equivalences describe the connection between convexity and positivity of the second derivative:

$$f \text{ convex} \leftrightarrow f'' \geq 0 \; \forall x$$
$$f \text{ strictly convex} \leftrightarrow f'' > 0 \; \forall x$$

These properties can be generalized to arbitrary dimensions using the properties of the Hessian matrix.

A2.3. Quadratic function

DEFINITION A2.6.– *A quadratic function is any function of the form:*

$$f(x) = \frac{1}{2}x^T A x - b^T x, \qquad \text{[A2.1]}$$

where $A \in \mathcal{M}_{nn}(\mathbb{R})$ *is a symmetric matrix and* $b \in \mathbb{R}^n$.

THEOREM A2.3.– A necessary and sufficient condition for the quadratic function defined by [A2.1] to be strictly convex is for A to be positive definite.

THEOREM A2.4.– The function $f\colon \mathbb{R}^n \to \mathbb{R}$ is convex (respectively, strictly convex) if and only if the Hessian matrix $\nabla^2 f(x)$ is positive semidefinite (respectively, positive definite) for any $x \in \mathbb{R}^n$.

Appendix 3

Optimization Toolbox

A3.1. Introduction

Matlab is a matrix-based language primarily intended for digital computer science. *Matlab* is specially designed for matrix computations, such as solving linear equations or factorizing matrices. Users who are not familiar with *Matlab* are invited to refer to [DAV 11, HAH 16]. Experienced users who wish to further optimize their code are invited to refer to [ALT 14].

Matlab has a series of built-in algorithms for finding roots (*root, fzero*), least squares (*lsqcurvefit, lsqlin*), the solution of systems of equations (*fsolve, fzero*) and minimization in one or several dimensions [RAD 18]. To minimize a function of one variable over a domain, we can use *fminbnd*. For functions of several variables, we can use *fminsearch*. For constrained problems, we can use *linprog* and *quadrprog* for the linear and quadratic cases, respectively. The function *fmincon* allows us to find the minimum of a constrained nonlinear multivariable problem.

Matlab includes an optimization toolbox for more complex problems, which automates the process of choosing the algorithm using a graphical user interface (GUI). *Matlab* uses multiple algorithms, depending on the problem being solved: "interior reflective Newton method", "trust-region-dogleg", "trust-region-reflective", "levenbergmarquardt", "simplex", "BFGS" and "MinMax". It includes four categories of solvers [MAT 17]:

1) minimizers: these are solvers that attempt to find a local minimum of the objective function. These solvers perform unconstrained optimization, linear programming, quadratic programming and nonlinear programming problems;

2) multiobjective minimizers: these are solvers that attempt to minimize the value of a set of functions, or find a position where a set of functions is below certain values;

3) equation solvers: these are solvers that attempt to find a solution to a nonlinear equation $f(x) = 0$;

4) least-squares solvers: solvers that attempt to minimize a sum of squares.

A3.2. Various functions

A complete list of the types of optimization problems supported by the solvers of the optimization toolbox is given in Table A3.1, together with the name of the relevant solver.

Type of problem	Solver	Description
Scalar optimization	*fminbnd*	Find the minimum of a function of one variable over a fixed interval
Unconstrained optimization	*fminunc* *fminsearch*	Find the minimum of an unconstrained multivariable problem (fminunc) using an optimization without derivatives (fminsearch)
Linear programs	*linprog*	Solve linear programs
Integer and mixed programming	*intlinprog*	Solve integer and mixed linear programs
Quadratic programming	*quadprog*	Solve quadratic programs
Constrained optimization	*fmincon*	Find the minimum of a constrained optimization problem
Minimax	*fminimax*	Solve problems with minimax constraints
Least squares problems	*lsqnonneg*	Solve least squares problems
Least squares with linear constraints	*lsqlin*	Solve least squares problems with linear constraints
Nonlinear least squares	*lsqnonlin*	Solve nonlinear least squares problems

Table A3.1. *Optimization problems supported by the optimization toolbox*

The optimization toolbox has four LP algorithms:

1) an interior point method;

2) a primal simplex algorithm;

3) a dual simplex algorithm;

4) an active set algorithm.

A3.3. *Matlab*'s optimization application

Matlab's optimization application (Figure A3.1) is an application built into the toolbox that simplifies the process of solving an optimization problem. It allows users to [MAT 17]:

- define the optimization problem;
- choose the solver and the solving method;
- solve the optimization problem;
- import and export an optimization problem, the solver options and the results;
- generate *Matlab* code to automate tasks.

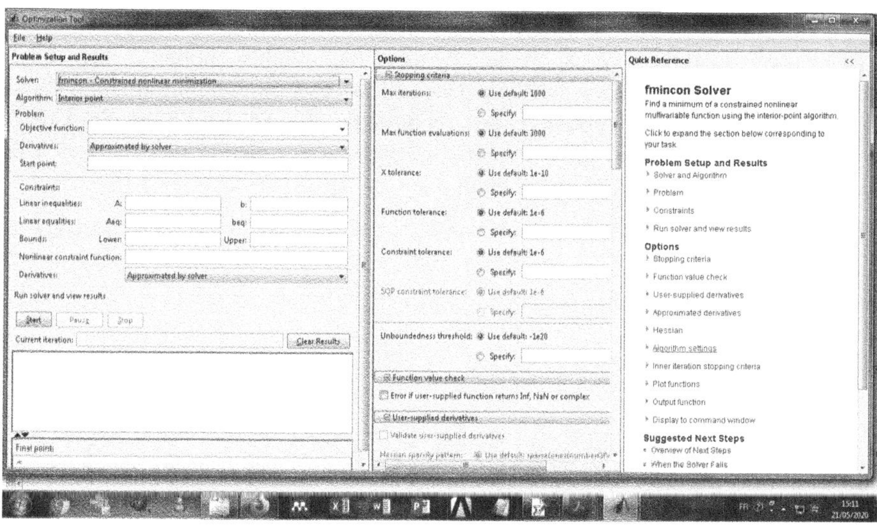

Figure A3.1. *Window that opens when calling optimtool. For a color version of this figure, see www.iste.co.uk/radi/optimizations.zip*

To open the optimization application, type *optimtool* into the command prompt, or launch it from the *Matlab* Apps tab (Figure A3.2).

Figure A3.2. *Applications in the Optimization Toolbox*

The user can select the solver and the specific algorithm to use in the "Problem configuration and results" tab. Every algorithm listed in Table A3.1 is available. Users can also select one of the four LP algorithms listed above, as well as entering the objective function, the linear equalities, linear inequalities and bounds of the LP problem in matrix form in the "Problem" and "Constraints" sections. For the active set algorithm, the user can either allow the algorithm to choose an initial point itself or specify a particular initial point in the "Start point" field. There are also other options in the "Options" tab that users can modify when solving LPs.

When a solver is running, the user can click on the "Pause" button to temporarily suspend the algorithm, or on the "Stop" button to stop the algorithm. When the solver terminates, "Run solver and display results" displays the results of the optimization process, including the value of the objective function, the solution of the LP problem and the reason why the algorithm terminated.

The user can import or export their work using suitable options in the "File" menu. The dialog box "Export to Workspace" allows users to send information about problems to the *Matlab* workspace as structures or objects. The "Generate Code" option in the "File" menu generates *Matlab* code as a function that the user can execute to solve the specified LP problem.

Appendix 4

Software

A4.1. Autonomous and multipurpose optimization software

1) *Matlab* optimization toolboxes. Matlab includes two optimization toolboxes that can solve various types of optimization problems, including linear, nonlinear, and multiobjective problems.

2) The XPRESS optimization algorithm suite is distributed by Dash Optimization [HRO 10]. The XPRESS MIP tool provides the ability to solve mixed programming problems using sophisticated implementations of branching and related algorithms, as well as cutting-plane algorithms. More information about the XPRESS suite is available on the website: www.fico.com/en/products/fico-xpress-optimization-suite.

3) CPLEX solvers. These have been commercially available since 1988, created by Robert E. Bixby, acquired by ILOG in 1997 and then by IBM in 2009. They are under proprietary license, but are free for students and teachers working in an academic setting. They can be used in C, C++, Java and Python. The solver uses a combination of algorithms and heuristics [PAR 10]. For more information about the free CPLEX solver, the user can visit the website: openopt.org/cplex.

4) Excel and Quattro Pro Solvers, developed by Frontline Systems [MOR 93], are available for solving small-scale integer programming problems using Excel worksheets [PRI 11]. The solving technique uses a branch-and-bound technique that applies a nonlinear method called GRG2 [FER 00].

5) The NEOS server is a free website hosted by the Argonne National Laboratory. The website offers several cutting-edge algorithms for optimization software, including for integer programming: www.neos-server.org/neos/solvers.

6) GAMS WORLD [VAN 07] is a website hosted by the international GAMS (General Algebraic Modeling System) community. It offers a large selection of source code for solving mixed integer nonlinear programs (MINLP), which are an important

and complex class of discrete optimization problems. Code for the MINLP solver can be found on the website: www.gamsworld.org/minlp/solvers.htm.

7) COIN-OR Linear Programming (CLP). Written in C++ in 2000, under EPL (Eclipse Public License), R, Java and Matlab interfaces are available.

8) GNU Linear Programming Kit (GLPK). Written in C, also in 2000, under GPLv3 (GNU Public License version 3).

A4.2. Packages for specific classes of problems

1) Constrained nonlinear problems are as follows:

a) NPSOL: NPSOL is a software program used to solve constrained optimization problems. It is especially effective with nonlinear problems whose functions and gradients are expensive to evaluate. The algorithms are numerically stable and guarantee global convergence. Details can be found on the website: www.sbsi-sol-optimize.com/asp/sol_product_npsol.htm.

b) GRG2: the GRG2 software [LAS 78] is used for nonlinear programming problems. It can be used as an autonomous system or a subprogram. It can manage up to 200 active constraints. Another implementation of GRG2, known as LSGRG2 [SMI 92], can solve large-scale optimization problems with more than 500 constraints.

2) Linear programming:

a) LSSOL: professional software for solving linear programs. The software claims to be numerically stable and is used to solve problems from statistics, economics and finance. Details are available on the website: www.sbsi-sol-optimize.com/asp/sol_product_lssol.htm.

b) CPLEX: CPLEX is used to solve linear programs and integer programs. It uses both the simplex algorithm and the interior-point algorithm for the linear case. It can be used directly or as a subprogram. The software claims to be robust and reliable with a pleasant interface. Details are available on the website: www.ibm.com/software/commerce/optimization/cplex-optimizer.

3) Unconstrained linear problems: the BTN software can be used to solve unconstrained nonlinear optimization problems, implemented within a parallel computing environment. This software is especially suitable for large-scale problems. Details are available in [NAS 92].

A4.3. Optimization software for design

The following is a representative list of different integration frameworks and their optimization functionality:

1) iSIGHT: this software helps engineers to manage the various programs required when executing simulation-based design processes, including commercial computer-assisted design/analysis software and Excel worksheets. iSIGHT allows these programs to be quickly integrated and automates their execution to accelerate the evaluation of multiple design alternatives. iSIGHT has multiobjective optimization features and contains cutting-edge multiobjective genetic algorithm routines. A toolbox is available, allowing users to integrate their own optimization algorithms, proprietary or otherwise. Because of a standard procedure, these algorithms can be made available through the iSIGHT optimization interface, including access to the specific configuration parameters of the algorithm. Details are available on the website: www.simulia.com.

2) PHX ModelCenter: Phoenix Integration was developed the PHX ModelCenter for process integration and design optimization. PHX ModelCenter is a process integration environment that supports the design team. It also allows multidimensional design spaces to be visualized to find the best designs. It allows users to construct design processes as a series of applications linked with a simple interface. Upon completion, users will have organized analysis models that make optimization and rapid exploration of the design space easier. More details are available on the website: www.phoenix-int.com.

3) LMS Samtech Boss Quattro: this is an application manager that works in a multidisciplinary environment and allows the user to explore the design space using optimization. It allows interfacing with computer-assisted design and analysis, internal code and management of technical sheets such as MSEXCEL. Boss Quattro is a professional software program by SIEMENS. It offers different software solutions that can be adapted to different applications and industrial environments. Details are available on the website: www.plm.automation.siemens.com.

4) modeFRONTIER: modeFRONTIER is an environment with a comprehensive and pleasant design that makes it easier to integrate multiple computer-assisted technologies (CAX), whether commercial or internal. The software has the ability to integrate structural analysis software based on the finite element method and numerical fluid mechanics. It also offers multiple optimization features, including gradient-based optimization, genetic algorithms and multiobjective optimization. The modeFRONTIER software can be seen as a wrapper around CAX tools that performs optimization by modifying the input variables and monitoring the output variables. Details are available on the website: www.esteco.com.

A4.4. Solvers for stochastic optimization

1) FortSP: a software program used to solve stochastic programming (SP) problems. It solves problems based on SP scenarios with recourse, as well as problems with chance constraints and integrated random constraints.

FortSP is available as a standalone executable that accepts inputs in the SMPS format and a library with an interface based on the C programming language. For multistep problems, the problem is reformulated into a deterministic problem. It also has an implementation of a cutting-plane algorithm for integrated random constraints. FortSP supports external linear programming applications such as CPLEX, FortMP and Gurboi because of library interfaces or nl files. These solvers are used to optimize the deterministic equivalent problem and subproblems in decomposition methods.

2) GAMS: high-level modeling system for mathematical programming and optimization. It consists of a language compiler and a high-performance solver. GAMS is designed for complex models with large full-scale modeling applications and allows users to construct large maintainable models that can quickly be adapted to new situations: www.gams.com.

3) COIN-OR: COmputational INfrastructure for Operations Research: this is an initiative for stimulating the development of open-source software for the operations research community: www.coin.or.org.

4) ORATIO: Another Mixed Logit Estimation Tool (Amlet) can be used for various operations research problems by taking advantage of the functions provided in its library. Amlet currently consists of three distinct blocks (www.slashbin.net/amlet):

- oratio;
- Ophelia;
- Amlet.

References

[ALT 14] ALTMAN Y.M., *Accelerating Matlab Performance: 1001 Tips to Speed Up Matlab Programs*, CRC Press, Boca Raton, FL, 2014.

[ANC 19] ANCU M., *Practical Optimization with MATLAB*, Cambridge Scholars Publishing, Cambridge, 2019.

[APP 06] APPLEGATE D.L., BIXBY R.E., CHVATAL V., COOK W.J., *The Traveling Salesman Problem: A Computational Study*, Princeton University Press, Princeton, 2006.

[ARO 98] ARORA S., "Polynomial time approximation schemes for Euclidean traveling salesman and other geometric problems", *Journal of the ACM*, vol. 45, no. 5, pp. 753–782, 1998.

[ASC 01] ASCHEUER N.F.M., GRÖTSCHEL M., "Solving the asymmetric travelling salesman problem with time windows by branch-and-cut", *Mathematical Programming*, vol. 90, no. 3, pp. 475–506, 2001.

[AVR 03] AVRIEL M., *Nonlinear Programming: Analysis and Methods*, Courier Corporation, North Chelmsford, MA, 2003.

[BAR 94] BARLES G., *Solutions de viscosité des équations de Hamilton-Jacobi*, Springer-Verlag, Paris, 1994.

[BAR 04] BARTY K., Contributions à la discrétisation des contraintes de mesurabilité pour les problèmes d'optimisation stochastique, PhD Thesis, École Nationale des Ponts et Chaussées, Paris, 2004.

[BAZ 13] BAZARAA M.S., SHERALI H.D., SHETTY C.M., *Nonlinear Programming: Theory and Algorithms*, John Wiley & Sons, New York, NY, 2013.

[BEK 06] BEKTAS T., "The multiple traveling salesman problem: An overview of formulations and solution procedures", *Omega*, vol. 34, no. 3, pp. 209–219, 2006.

[BEL 56] BELLMAN R., "Dynamic programming and Lagrange multipliers", *Proceedings of the National Academy of Sciences of the United States of America*, vol. 42, no. 10, p. 767, 1956.

[BEN 62] BENDERS J., "Partioning procedures for solving mixed variables programming problems", *Numerische Mathematic*, vol. 4, pp. 238–252, 1962.

[BER 95] BERTSEKAS D., *Dynamic Programming and Optimal Control*, vol. 1, Athena Scientific, Belmont, MA, 1995.

[BER 99] BERTSEKAS D., *Nonlinear Programming*, Athena Scientific, Belmont, MA, 1999.

[BER 00] BERTSEKAS D.P., TSITSIKLIS J.N., "Gradient convergence in gradient methods", *SIAM, Journal of Optimization*, vol. 10, no. 3, pp. 627–642, 2000.

[BER 01] BERGOUNIOUX M., *Optimisation et contrôle des systèmes linéaires*, Dunod, Paris, 2001.

[BLA 77] BLAND R., "New finite pivoting rules for the simplex method", *Mathematics of Operations Research*, vol. 2, no. 2, pp. 103–107, 1977.

[BOL 00] BOLAND N.L., CLARKE -L.W., NEMHAUSER G.L., "The asymmetric traveling salesman problem with replenishment arcs", *European Journal of Operational Research*, vol. 123, no. 2, pp. 408–427, 2000.

[BON 06] BONNANS J.-F., GILBERT J.C., LEMARÉCHAL C., SAGASTIZÁBAL C.A., *Numerical Optimization: Theoretical and Practical Aspects*, Springer Science & Business Media, Berlin, Heidelberg, 2006.

[CAR 07] CARLIER G., Programmation Dynamique, Course notes, ENSAE, Paris, 2007.

[CHA 96] CHAO M., GOLDEN B.L., WASIL E.A., "The team orienteering problem", *European Journal of Operational Research*, vol. 8, pp. 464–474, 1996.

[CHR 79] CHRISTOFIDES N., MINGOZZI A., TOTH P., "The vehicle routing problem", *Combinatorial Optimization*, vol. 11, pp. 315–338, 1979.

[CHV 83] CHVATÁL V., *Linear Programming*, Operations Research, W.H. Freeman, New York, NY, 1983.

[COH 90] COHEN G., CULIOLI J.-C., "Decomposition coordination algorithms for stochastic optimization", *SIAM Journal Control Optimization*, vol. 28, no. 6, pp. 1372–1403, 1990.

[COR 02] CORMEN T.H., LEISERSON C.E., RIVEST R.L., STEIN C., *Introduction à l'algorithmique*, Dunod, Paris, 2002.

[CRA 14] CRAINIC T., SEMET F., "Operations research and goods transportation", in PASCHOS, T.V. (ed.), *Applications of Combinatorial Optimization*, 2nd ed., ISTE Ltd, London, and John Wiley & Sons, New York, NY, 2014.

[DAN 54] DANTZIG G.B., FULKERSON D.R., JOHNSON S., "Solution of a large-scale traveling-salesman problem", *Journal of the Operations Research Society of America*, vol. 2, no. 4, pp. 393–410, 1954.

[DAN 59] DANTZIG G., RAMSER J., "The truck dispatching problem", *Management Science*, vol. 6, no. 1, pp. 80–91, 1959.

[DAV 11] DAVIS T.A., *Matlab Primer*, 8th ed., Chapman & Hall/CRC, Boca Raton, FL, 2011.

[DES 91] DESROCHERS M., LAPORTE G., "Improvements and extensions to the Miller-Tucker-Zemlin subtour elimination constraints", *Operations Research Letters*, vol. 10, no. 1, pp. 27–36, 1991.

[DES 06] DESAULNIERS G.D.J., SOLOMON M.M., *Column Generation*, vol. 5, Springer Science & Business Media, Berlin, Heidelberg, 2006.

[ELA 09] EL ALEM W.E.H.A., ELLAIA R., "Global optimization for structural design", *Revue de mécanique appliquée et théorique*, vol. 2, no. 1, pp. 125–136, 2009.

[ELH 10] KHARMANDA G., EL HAMI A., SOUZA DE CURSI E., "Reliability-based Design Optimization (RBDO)" in BREITKOPF C., COELHO R.F. (eds), *Multidisciplinary Design Optimization in Computational Mechanics*, ISTE Ltd, London, and John Wiley & Sons, New York, NY, 2010.

[ELH 11] EL HAMI A., RADI B., *Fiabilité et optimisation des systèmes : théorie et applications, cours et exercices corrigés*, Ellipses, Paris, 2011.

[ELH 13] EL HAMI A., RADI B., *Uncertainty and Optimization in Structural Mechanics*, John Wiley & Sons, New York, NY, 2013.

[ELM 10] EL MAGHRI M., RADI B., "Convex functionals with good asymptotic behavior on a subset", *Applied Mathematics E-Notes*, vol. 10, pp. 52–60, 2010.

[FEI 01] FEILLET D., Problèmes de tournées avec gains : étude et application au transport inter-usines, PhD Thesis, École Centrale de Paris, Paris, 2001.

[FER 00] FERRIS M.C., MESNIER M.P., "NEOS and CONDOR: Solving optimization problems over the Internet", *ACM Transactions on Mathematical Software*, vol. 26, no. 1, pp. 1–18, 2000.

[FIS 94] FISCHETTI M.T.P., VIGO D., "A branch-and-bound algorithm for the capacitated vehicle routing problem on directed graphs", *Operations Research*, vol. 42, no. 5, pp. 846–859, 1994.

[FLE 10] FLEURY G., LACOMME P., *Programmation linéaire avancée*, Ellipses, Paris, 2010.

[GAS 75] GASS S., *Linear Programming*, McGraw Hill, New York, NY, 1975.

[GHE 07] GHEDIRA K., *Optimisation combinatoire par métaheuristiques*, Technip, Paris, 2007.

[GOL 87] GOLDEN B., LEVY L., VOHRA R.V., "The orienteering problem", *Naval Research Logistics*, vol. 34, pp. 307–318, 1987.

[GOM 63] GOMORV R., "An algorithm for integer solutions to linear programs", in GRAVES R.L., WOLFE P. (eds), *Recent Advances in Mathematical Programming*, McGraw-Hill, New York, NY, 1963.

[GOU 06] GOUVEIA L., PESNEAU P., "On extended formulations for the precedence constrained asymmetric traveling salesman problem", *Networks: An International Journal*, vol. 48, no. 2, pp. 77–89, 2006.

[GUT 06] GUTIN G., PUNNEN A.P., *The Traveling Salesman Problem and Its Variations*, vol. 12, Springer Science & Business Media, Berlin, Heidelberg, 2006.

[HAH 16] HAHN B., VALENTINE D.T., *Essential Matlab for Engineers and Scientists*, 5th ed., Academic Press, Cambridge, MA, 2016.

[HAJ 03] HAJJI O., Contributions au développement de méthodes stochastiques. Application à la conception des dispositifs électroniques, PhD Thesis, ENIM, Monastir, 2003.

[HEI 03] HEITSCH H.R.W., RÖMISCH W., "Scenario reduction algorithms in stochastic programming", *Computational Optimization and Applications*, vol. 24, pp. 187–206, 2003.

[HEI 06] HEITSCH H.R.W., STRUGAREK C., "Stability of multistage stochastic programs", *SIAM Journal on Optimization*, vol. 17, pp. 511–525, 2006.

[HIL 90] HILLIER F., LIEBERMAN G., *Introduction to Operations Research*, McGraw-Hill, New York, NY, 1990.

[HRO 10] HROMKOVIC J., *Algorithmics for Hard Problems: Introduction to Combinatorial Optimization, Randomization, Approximation, and Heuristics*, Springer, Berlin, Heidelberg, 2010.

[JEN n/a] JENSEN P., Benders' method, available at: https://www.me.utexas.edu/~jensen/ORMM/methods/unit/integer/subunits/teach_int/benders.pdf.

[JÜN 95] JÜNGER M.R.G., RINALDI G., "The traveling salesman problem", *Handbooks in Operations Research and Management Science*, vol. 7, pp. 225–330, 1995.

[KAL 94] KALL P., STEIN W., *Stochastic Programming*, John Wiley & Sons, New York, NY, 1994.

[KAN 92] KANTOR M., ROSENWEIN M., "The team orienteering problem", *European Journal of Operational Research*, vol. 43, pp. 629–635, 1992.

[KAR 84] KARMARKAR N., "A new polynomial time algorithm for linear programming", *Combinatorica*, vol. 4, pp. 373–395, 1984.

[KHA 79] KHACHIAN L., "A polynomial algotithm for linear programming", *Soviet Mathematics Doklady*, vol. 20, pp. 191–194, 1979.

[KLE 72] KLEE V., MINTY G., "How good is the simplex algorithm?", in SHISHA O. (ed.), *Inequalities III*, Academic Press, New York, NY, 1972.

[KNU 71] KNUTH D.E., "Optimum binary search trees", *Acta Informatica*, vol. 1, no. 1, pp. 14–25, 1971.

[KOR 11] KORTE B., VYGEN J., *Combinatorial Optimization, Theory and Algorithms*, vol. 21, Springer, Berlin, Heidelberg, 2011.

[LAP 92] LAPORTE G.L.F., MERCURE H., "The vehicle routing problem with stochastic travel times", *Transportation Science*, vol. 26, pp. 161–170, 1992.

[LAS 78] LASDON L.S., WAREN A.D., JAIN A., RATNER M., "Design and testing of a generalized reduced gradient code for nonlinear programming", *ACM Transactions on Mathematical Software*, vol. 4, pp. 34–50, 1978.

[LAW 85] LAWLER E.L., *The Traveling Salesman Problem: A Guided Tour of Combinatorial Optimization*, Wiley, Chichester, 1985.

[LEI 01] LEISERSON C.E., RIVEST R.L., STEIN C., CORMEN T.H., *Introduction to Algorithms*, The MIT Press, Cambridge, MA, 2001.

[LEN 81] LENSTRA J., RINNOOY KAN A., "Complexity of the vehicle routing and scheduling problems", *Networks*, vol. 11, pp. 221–228, 1981.

[LUC 89] LUCAS R.E., STOKEY N., *Recursive Methods in Economics Dynamics*, Harvard University Press, Cambridge, MA, 1989.

[MAC 83] MACULAN N., "Relaxation lagrangienne : le problème du knapsack 0–1", *INFOR: Information Systems and Operational Research*, vol. 21, no. 4, pp. 315–327, 1983.

[MAK 08] MAKRIZI A., RADI B., EL HAMI A., "Solution of the topolgy optimization problem based subdomain method", *Applied Mathematical Sciences*, vol. 2, no. 42, pp. 2047–2062, 2008.

[MAT 17] MATHWORKS, *Optimization Toolbox: User's Guide*, The MathWorks, Inc., 2017.

[MIL 60] MILLER C.E., TUCKER A.W., ZEMLIN R.A., "Integer programming formulation of traveling salesman problems", *Journal of the ACM*, vol. 7, no. 4, pp. 326–329, 1960.

[MOR 93] MORÉ J.J., WRIGHT S.J., *Optimization Software Guide*, Society for Industrial and Applied Mathematics, Philadelphia, PA, 1993.

[NAB 07] NABAA M., ZEDDINI B., TRANOUEZ P., "Approche décentralisée pour résoudre le problème du transport à la demande", *Schedae, prépublication*, no. 23, pp. 133–140, 2007.

[NAR 77] NARENDRA P.M., FUKUNAGA K., "A branch and bound algorithm for feature subset selection", *Transactions on Computers*, no. 9, pp. 917–922, 1977.

[NAS 92] NASH S.G., SOFER A., "BTN: Software for parallel and unconstrained optimization", *ACM Transactions on Mathematical Software*, vol. 18, pp. 414–448, 1992.

[NEL 65] NELDER J., MEAD R., "A simplex method for function minimization", *Computer Journal*, vol. 7, no. 4, pp. 308–313, 1965.

[NEM 98] NEMHAUSER G., WOLSEY L., *Integer and Combinatorial Optimzation*, Wiley, New York, NY, 1998.

[NOC 99] NOCEDAL J., WRIGHT S.J., *Numerical Optimization*, Springer, Berlin, Heidelberg, 1999.

[ÖNC 09] ÖNCAN T., ALTNEL K.I., LAPORTE G., "A comparative analysis of several asymmetric traveling salesman problem formulations", *Computers & Operations Research*, vol. 36, no. 3, pp. 637–654, 2009.

[PAD 91] PADBERG M., RINALDI G., "A branch-and-cut algorithm for the resolution of large-scale symmetric traveling salesman problems", *SIAM Review*, vol. 33, no. 1, pp. 60–100, 1991.

[PAL 04] PALETTA G., TRIKI C., "Solving the asymmetric traveling salesman problem with periodic constraints", *Networks: An International Journal*, vol. 44, no. 1, pp. 31–37, 2004.

[PAP 77] PAPADIMITRIOU C.H., "The Euclidean traveling salesman problem is NP-complete", *Theoretical Computer Science*, vol. 4, no. 3, pp. 237–244, 1977.

[PAR 10] PARDALOS P.M., ZHENG Q.P., ARULSELVAN A., *Deterministic Global Optimization*, Encyclopedia of Operations Research and Management Science, Wiley, New York, NY, 2010.

[PER 15] PEREIRA D.L., GENDREAU M., SALLES DA CUNHA A., "Branch-and-cut and branch-and-cut-and-price algorithms for the adjacent only quadratic minimum spanning tree problem", *Networks*, vol. 65, no. 4, pp. 367–379, 2015.

[PFL 01] PFLUG G.C., "Scenario tree generation for multiperiod financial optimization by optimal discretization", *Mathematical Programming*, no. 89, pp. 251–271, 2001.

[PRI 11] PRINS C., SEVAUX M., *Programmation linéaire avec Excel*, Eyrolles, Paris, 2011.

[RAD 11] RADI B., AYADI A., CHEROUAT A., GIRAUD-MOREAU L., EL HAMI A., "Identification of the characteristics of hydroformed structures using optimization methods", *Key Engineering Materials*, vol. 473, no. 62, pp. 723–730, 2011.

[RAD 16] RADI B., EL HAMI A., *Material Forming Processes*, ISTE Ltd, London, and John Wiley & Sons, New York, NY, 2016.

[RAD 18] RADI B., EL HAMI A., *Advanced Numerical Methods with Matlab® 1*, ISTE Ltd, London, and John Wiley & Sons, New York, NY, 2018.

[ROB 51] ROBBINS H., MONRO S., "A stochastic approximation method", *Annals of Mathematical Statistics*, vol. 22, pp. 400–407, 1951.

[SAL 75] SALKIN H., *Integer Programming*, Addison-Wesley, Boston, MA, 1975.

[SAR 05] SARIN S.C., SHERALI H.D., BHOOTRA A., "New tighter polynomial length formulations for the asymmetric traveling salesman problem with and without precedence constraints", *Operations Research Letters*, vol. 33, no. 1, pp. 62–70, 2005.

[SAV 95] SAVELSBERGH M., "Local search for routing problems with time windows", *Annals of Operations Research*, vol. 4, pp. 285–305, 1995.

[SMA 19] SMAOUI H., MAQSOUD A., KAIDI S., "Transmissivity identification by combination of CVFEM and genetic algorithm: Application to the coastal aquifer", *Mathematical Problems in Engineering*, available at: https://doi.org/10.1155/2019/3463607, 2019.

[SMI 92] SMITH S., LASDON L.S., "Solving large sparse nonlinear programs using GRG", *ORSA Journal on Computing*, vol. 4, pp. 1–15, 1992.

[SNI 06] SNIEDOVICH M., LEW A., "Dynamic programming: An overview", *Control and Cybernetics*, vol. 35, no. 3, pp. 5-13, 2006.

[TEG 12] TEGHEM J., *Recherche opérationnelle*, vol. 1, Ellipses, Paris, 2012.

[TOT 01] TOTH P., VIGO D., *The Vehicle Routing Problem*, SIAM, Philadelphia, PA, 2001.

[TSI 84] TSILIGIRIDES L., "Heuristics methods applied to orienteering", *Journal of the Operational Research Society*, vol. 35, pp. 797–809, 1984.

[VAN 96] VANDERBECK F., WOLSEY L.A., "An exact algorithm for IP column generation", *Operations Research Letters*, vol. 19, no. 4, pp. 151–159, 1996.

[VAN 07] VANDERPLAATS G.N., *Multidiscipline Design Optimization*, Vanderplaats Research and Development, Inc., Colorado Springs, CO, 2007.

[WOL 98] WOLSEY L., *Integer Programming*, John WIley & Sons, New York, NY, 1998.

Index

Other titles from

ISTE

in

Mechanical Engineering and Solid Mechanics

2021

DAHOO Pierre Richard, POUGNET Philippe, EL HAMI Abdelkhalak
Applications and Metrology at Nanometer Scale 1: Smart Materials, Electromagnetic Waves and Uncertainties
(Reliability of Multiphysical Systems Set – Volume 9)

2020

SALENÇON Jean
Elastoplastic Modeling

2019

BAYLE Franck
Reliability of Maintained Systems Subjected to Wear Failure Mechanisms: Theory and Applications
(Reliability of Multiphysical Systems Set – Volume 8)

BEN KAHLA Rabeb, BARKAOUI Abdelwahed, MERZOUKI Tarek
Finite Element Method and Medical Imaging Techniques in Bone Biomechanics
(Mathematical and Mechanical Engineering Set – Volume 8)

IONESCU Ioan R., QUEYREAU Sylvain, PICU Catalin R., SALMAN Oguz Umut
Mechanics and Physics of Solids at Micro- and Nano-Scales

LE VAN Anh, BOUZIDI Rabah
Lagrangian Mechanics: An Advanced Analytical Approach

MICHELITSCH Thomas, PÉREZ RIASCOS Alejandro, COLLET Bernard,
NOWAKOWSKI Andrzej, NICOLLEAU Franck
Fractional Dynamics on Networks and Lattices

SALENÇON Jean
Viscoelastic Modeling for Structural Analysis

VÉNIZÉLOS Georges, EL HAMI Abdelkhalak
Movement Equations 5: Dynamics of a Set of Solids
(Non-deformable Solid Mechanics Set – Volume 5)

2018

BOREL Michel, VÉNIZÉLOS Georges
Movement Equations 4: Equilibriums and Small Movements
(Non-deformable Solid Mechanics Set – Volume 4)

FROSSARD Etienne
Granular Geomaterials Dissipative Mechanics: Theory and Applications in Civil Engineering

RADI Bouchaib, EL HAMI Abdelkhalak
Advanced Numerical Methods with Matlab® 1: Function Approximation and System Resolution
(Mathematical and Mechanical Engineering SET – Volume 6)
Advanced Numerical Methods with Matlab® 2: Resolution of Nonlinear, Differential and Partial Differential Equations
(Mathematical and Mechanical Engineering SET – Volume 7)

SALENÇON Jean
Virtual Work Approach to Mechanical Modeling

2017

BOREL Michel, VÉNIZÉLOS Georges
Movement Equations 2: Mathematical and Methodological Supplements (Non-deformable Solid Mechanics Set – Volume 2)
Movement Equations 3: Dynamics and Fundamental Principle (Non-deformable Solid Mechanics Set – Volume 3)

BOUVET Christophe
Mechanics of Aeronautical Solids, Materials and Structures
Mechanics of Aeronautical Composite Materials

BRANCHERIE Delphine, FEISSEL Pierre, BOUVIER Salima, IBRAHIMBEGOVIC Adnan
From Microstructure Investigations to Multiscale Modeling: Bridging the Gap

CHEBEL-MORELLO Brigitte, NICOD Jean-Marc, VARNIER Christophe
From Prognostics and Health Systems Management to Predictive Maintenance 2: Knowledge, Traceability and Decision (Reliability of Multiphysical Systems Set – Volume 7)

EL HAMI Abdelkhalak, RADI Bouchaib
Dynamics of Large Structures and Inverse Problems (Mathematical and Mechanical Engineering Set – Volume 5)
Fluid-Structure Interactions and Uncertainties: Ansys and Fluent Tools (Reliability of Multiphysical Systems Set – Volume 6)

KHARMANDA Ghias, EL HAMI Abdelkhalak
Biomechanics: Optimization, Uncertainties and Reliability (Reliability of Multiphysical Systems Set – Volume 5)

LEDOUX Michel, EL HAMI Abdelkhalak
Compressible Flow Propulsion and Digital Approaches in Fluid Mechanics (Mathematical and Mechanical Engineering Set – Volume 4)
Fluid Mechanics: Analytical Methods (Mathematical and Mechanical Engineering Set – Volume 3)

MORI Yvon
Mechanical Vibrations: Applications to Equipment

2016

BOREL Michel, VÉNIZÉLOS Georges
Movement Equations 1: Location, Kinematics and Kinetics
(Non-deformable Solid Mechanics Set – Volume 1)

BOYARD Nicolas
Heat Transfer in Polymer Composite Materials

CARDON Alain, ITMI Mhamed
New Autonomous Systems
(Reliability of Multiphysical Systems Set – Volume 1)

DAHOO Pierre Richard, POUGNET Philippe, EL HAMI Abdelkhalak
Nanometer-scale Defect Detection Using Polarized Light
(Reliability of Multiphysical Systems Set – Volume 2)

DE SAXCÉ Géry, VALLÉE Claude
Galilean Mechanics and Thermodynamics of Continua

DORMIEUX Luc, KONDO Djimédo
Micromechanics of Fracture and Damage
(Micromechanics Set – Volume 1)

EL HAMI Abdelkhalak, RADI Bouchaib
Stochastic Dynamics of Structures
(Mathematical and Mechanical Engineering Set – Volume 2)

GOURIVEAU Rafael, MEDJAHER Kamal, ZERHOUNI Noureddine
From Prognostics and Health Systems Management to Predictive Maintenance 1: Monitoring and Prognostics
(Reliability of Multiphysical Systems Set – Volume 4)

KHARMANDA Ghias, EL HAMI Abdelkhalak
Reliability in Biomechanics
(Reliability of Multiphysical Systems Set –Volume 3)

MOLIMARD Jérôme
Experimental Mechanics of Solids and Structures

RADI Bouchaib, EL HAMI Abdelkhalak
Material Forming Processes: Simulation, Drawing, Hydroforming and Additive Manufacturing
(Mathematical and Mechanical Engineering Set – Volume 1)

2015

KARLIČIĆ Danilo, MURMU Tony, ADHIKARI Sondipon, MCCARTHY Michael
Non-local Structural Mechanics

SAB Karam, LEBÉE Arthur
Homogenization of Heterogeneous Thin and Thick Plates

2014

ATANACKOVIC M. Teodor, PILIPOVIC Stevan, STANKOVIC Bogoljub, ZORICA Dusan
Fractional Calculus with Applications in Mechanics: Vibrations and Diffusion Processes
Fractional Calculus with Applications in Mechanics: Wave Propagation, Impact and Variational Principles

CIBLAC Thierry, MOREL Jean-Claude
Sustainable Masonry: Stability and Behavior of Structures

ILANKO Sinniah, MONTERRUBIO Luis E., MOCHIDA Yusuke
The Rayleigh–Ritz Method for Structural Analysis

LALANNE Christian
Mechanical Vibration and Shock Analysis – 5-volume series – 3rd edition
Sinusoidal Vibration – Volume 1
Mechanical Shock – Volume 2
Random Vibration – Volume 3
Fatigue Damage – Volume 4
Specification Development – Volume 5

LEMAIRE Maurice
Uncertainty and Mechanics

2013

ADHIKARI Sondipon
Structural Dynamic Analysis with Generalized Damping Models: Analysis

ADHIKARI Sondipon
Structural Dynamic Analysis with Generalized Damping Models: Identification

BAILLY Patrice
Materials and Structures under Shock and Impact

BASTIEN Jérôme, BERNARDIN Frédéric, LAMARQUE Claude-Henri
Non-smooth Deterministic or Stochastic Discrete Dynamical Systems: Applications to Models with Friction or Impact

EL HAMI Abdelkhalak, RADI Bouchaib
Uncertainty and Optimization in Structural Mechanics

KIRILLOV Oleg N., PELINOVSKY Dmitry E.
Nonlinear Physical Systems: Spectral Analysis, Stability and Bifurcations

LUONGO Angelo, ZULLI Daniele
Mathematical Models of Beams and Cables

SALENÇON Jean
Yield Design

2012

DAVIM J. Paulo
Mechanical Engineering Education

DUPEUX Michel, BRACCINI Muriel
Mechanics of Solid Interfaces

ELISHAKOFF Isaac *et al.*
Carbon Nanotubes and Nanosensors: Vibration, Buckling and Ballistic Impact

GRÉDIAC Michel, HILD François
Full-Field Measurements and Identification in Solid Mechanics

GROUS Ammar
Fracture Mechanics – 3-volume series
Analysis of Reliability and Quality Control – Volume 1
Applied Reliability – Volume 2
Applied Quality Control – Volume 3

RECHO Naman
Fracture Mechanics and Crack Growth

2011

KRYSINSKI Tomasz, MALBURET François
Mechanical Instability

SOUSTELLE Michel
An Introduction to Chemical Kinetics

2010

BREITKOPF Piotr, FILOMENO COELHO Rajan
Multidisciplinary Design Optimization in Computational Mechanics

DAVIM J. Paulo
Biotribolgy

PAULTRE Patrick
Dynamics of Structures

SOUSTELLE Michel
Handbook of Heterogenous Kinetics

2009

BERLIOZ Alain, TROMPETTE Philippe
Solid Mechanics using the Finite Element Method

LEMAIRE Maurice
Structural Reliability

2007

GIRARD Alain, ROY Nicolas
Structural Dynamics in Industry

GUINEBRETIÈRE René
X-ray Diffraction by Polycrystalline Materials

KRYSINSKI Tomasz, MALBURET François
Mechanical Vibrations

KUNDU Tribikram
Advanced Ultrasonic Methods for Material and Structure Inspection

SIH George C. *et al.*
Particle and Continuum Aspects of Mesomechanics

Printed and bound by CPI Group (UK) Ltd, Croydon, CR0 4YY

13/09/2026

14961021-0005